Attack of the Phasmii

Curse of Rogares

By: K Clay Dawson

Copyright

Dedication

I dedicate this book, "Attack of the Phasmii" to my loving, devoted wife and greatest nurse ever, Patricia Ann Smith.

About the author

Hello! My name is K Clay Dawson. I'm an author and retired trial lawyer. My interests include sports, writing, and riding four-wheelers. I've tested ATV trails from Half-Hell Swamp, North Carolina, to Rubicon Springs, California.

I began the practice of law in 1982 and retired in 2022. As an attorney, I had the honor and privilege of representing thousands of people through difficult times. However, whatever they suffered, I tended to suffer empathically, so practicing law was often challenging. In one case, I slept in a motel room in northern Georgia, facing the door with a loaded .45 in my hand.

I fought for my clients. Some judges didn't like that. I was threatened with contempt more than once. I told my wife not to bond me out of jail if one of them held me in contempt. She objected, but I explained that the State Bar would investigate, and Judges don't want that. Besides, being incarcerated would allow more time to write.

Retirement allowed me to complete my books. Most are sci-fi, which I've loved since I was a kid. Ray

Harryhausen's animation opened time portals and galaxies to me. There were great movies, including "Forbidden Planet," "Twenty Million Miles to Earth," and "The Day the Earth Stood Still," *Klaatu barada nikto*! My father brought home the movie poster for "The Valley of Gwangi."

So, why do I write? For one thing, writing comes easily to me. All I need to do is type, then Characters and stories come to life.

Please enjoy "Attack of the Phasmii."

Table of Contents

COPYRIGHT 2

DEDICATION 3

ABOUT THE AUTHOR 4

PREFACE 10

CHAPTER ONE 13

Attack of the Phasmii 13

CHAPTER TWO 42

Attack on Outpost Semotus 42

CHAPTER THREE 75

Promises of Unwounded Blood 75

CHAPTER FOUR 95

Astraea's Promise 95

CHAPTER FIVE 110

Call to Duty 110

CHAPTER SIX 114

Faster Than They Look 114

CHAPTER SEVEN 119

Captain On Board 119

CHAPTER EIGHT 147

Frugie's Fight 147

Journey Past Hades 165

CHAPTER TEN 175

FOURTEEN MINUTES 175

CHAPTER ELEVEN 203

DOCTOR VARUS'S DEPARTURE 203

CHAPTER TWELVE 219

Funeral Barge 219

CHAPTER THIRTEEN 230

Joker Squadron 230

CHAPTER FOURTEEN 252

Conspiracy of the Humiliated 252

CHAPTER FIFTEEN 263

The Naval Inquiry Begins 263

CHAPTER SIXTEEN 286

After Morning Recess 286

CHAPTER SEVENTEEN 291

Secret Weapon 291

CHAPTER EIGHTEEN 307

Testimony of City Councilman Fidelis 307

CHAPTER NINETEEN 312

Another Fourteen Minutes 312

CHAPTER TWENTY 336

Captain Astraea Meets Prime Minister Morus ... 336

CHAPTER TWENTY-ONE 353

Blood Alley – Prelude to Comet Day 353

CHAPTER TWENTY-TWO 363

Dawn of Comet Day 363

PREFACE

FOR BOOK ONE – CURSE OF ROGARES

Earth has a twin named Gemina. Scientists suggest the planets are "quantum entanglements." That's a phenomenon in which two separate, identical particles move and act the same way simultaneously. No one understands how or why.

Scientists struggle with the concept of quantum mechanics. Albert Einstein wasted much of his life trying to disprove it. Yet, quantum physics was used to create transistors, micro-transistors, and modern technology. No one understands why quantum physics works. However, modern science accepts it as true. Ironically, today's advanced science is based on faith.

Gemina is identical to Earth in mass, size, and atmosphere. Each planet has the same continents, oceans, rivers, and lakes. They shared a common history until 264 BC. On Earth, the Republic of Rome and the Republic of Carthage went to war against each other. After three Punic Wars, Carthage was destroyed. Its advanced, inclusive culture was quashed. Rome's promising Republic devolved into centuries of tyranny, dictatorship, and constant war.

Other great cultures perished during the tumultuous Age of Caesars. Their demise was followed by Dark Ages, plagues, and intolerance. Earth suffered twenty-three centuries of suppression and hostility.

Gemina escaped that hell. On Gemina, the two mighty Republics entered into a prosperous alliance. Roman genius and Carthaginian commerce flourished together. An enlightened Mediterranean civilization spread across the world. Unlike its quantum twin Earth, music and art weren't needed on Gemina to escape repression. Instead, they had practical applications.

Geminum exceeded its Terran counterparts in technology. They had pulse and graviton technology. However, Gemina lacked Earth's mastery of arts and music.

Unknown to either planet, they shared a common threat. A hostile empire intended to attack both planets and enslave their survivors. They were a clever, merciless enemy that had already built a vast empire throughout a distant system. Gemina was their immediate target.

[End of Preface]

CHAPTER ONE

Attack of the Phasmii

Six bells welcomed a new morning aboard the *GSS Lioness.* Gemina's finest interstellar Destroyer raced through deep space. Aboard her, young Ensign Vox Sonos slowly awoke. Time for dreams had ended. Something dark approached.

Cotton linens clung tightly to the young Ensign. She drew them towards her shoulder, which uncovered the smaller three toes of her left foot. As was customary of women from Outpost Semotus, her toenails were polished bright pink. Amused by her discovery, the young Ensign pulled back her blanket.

An impressive, twining tattoo coiled from her left ankle. A jade viper's tail twined tightly around her calf, past her knee. Sonos cast aside her sheets to see the serpent coiled tightly up and around her thigh. Its triangular head nestled patiently. She was proud of her tattoo.

An inspired tattooist on Semotus had artfully crafted Sonos' s magnificent tat. He was known only as "Razzouk,[1]" the man who paints life onto hungry flesh. He inked the snake in vibrant graphene colors. Its stunning image craftily created a second serpent from the young woman's smooth, uncolored thew. Two night-adders twined the ensign's leg into a Caduceus.

Her body art displayed passion and a tremendous pain threshold. She endured twenty-two hours of scratching and burning needles as Razzouk completed his work. Days of denied suffering followed.

Razzouk's subdermal rendition was stunning, masterly. Yet, few saw it.

There was one young man whom Sonos favored. His name was Callistus Caritas. Ensign Sonos' handsome suitor was an assistant director of communications at Outpost Semotus. He and Sonos had grown close in recent months. They were falling in love.

---

[1] There was a famous tattooist named Razzouk on the Twin Planet Earth, *circa* 1400 AD. The artist's family continues his traditions to this date in Jerusalem.

The young Ensign rose from her bunk. White cotton linen slipped slowly from her shoulders to the deck. She strode, undressed, and barefoot to the shower. Within minutes, she was uniformed and ready for duty.

Ensign Sonos hurried to the bridge. Her Commander, Jedrick Aegus Clades, welcomed her to the bridge. He was a tall, attractive man, with a pleasing voice and commanding presence, but Clades was easily old enough to be her father. Their relationship was one of mutual respect. He was the ship's captain. She was an able officer under his command. Sonos took her station.

Her equipment flashed red and white collision alarms. Klaxons sounded throughout the ship. Black shadows raced past the *Lioness*.

Ensign Sonos loudly reported, "Commander, six vessels just crossed our bow."

"Identify them, Miss Sonos."

"They appear to be destroyers. I've never seen any like them, sir."

"Hail them, Ensign."

"Sir, they may be hostile. A hail could reveal our position."

"Agreed, Ensign. Bounce our hail off every moon, asteroid, or other space rock near them. They won't know where we are."

"Aye, sir. Unidentified vessels, this is the *GSS Lioness*. Please identify yourself."

"Instruct them to state their purpose, Ensign."

"Commander, they're ignoring our hail."

"What is their course?"

"They're headed toward Outpost Semotus. It doesn't look good, sir."

Other than deep space freighters and commerce vessels, few ships port at Semotus. Why was an unidentified squadron of warships enroute to the distant Outpost without notice or authorization?

"Sir, another six warships just passed our bow. They are of like kind."

Commander Clades was concerned. "Ensign Sonos, tell Lieutenant Commander Paraclitis that twelve unidentified warships passed us, and they are headed towards Outpost Semotus. Have Mr. Paraclitis assemble all department heads in the ready room. We have a situation."

After hailing the Lieutenant Commander, Miss Sonos reported, "Mr. Paraclitis is gathering the ship's officers."

Lieutenant Commander *Titus Leo Paraclitis* was Clade's trusted Executive Officer. The two men shared history. They shed blood together during the Barbary Wars. Their bond was unbreakable.

He was popular among his fellow officers and crew. The Lieutenant Commander had been promoted from the ranks. His father was a former chief petty officer. Mr. Paraclitis understood and appreciated common Stellarii[2] and officers alike. He had earned their trust and respect.

Paraclitis also wore impressive tattoos. Each revealed an aspect of his character. During Officer

---

[2] The Geminum term for star sailors.

Training School, Paraclitis had a fierce, imperial eagle inked upon his left bicep. The screeching raptor grasped an anchor in its talons. It demonstrated his devotion to the Navy.

Years later, he added another tat. It was quite different. Unlike the bird of prey, the Commander's second tattoo professed love.

Her name was Paloma. They met on Outpost Semotus and quickly were enamored of one another. The morning that his second tattoo was inked, they awoke together beneath satin sheets. He embraced her warm figure.

"Good morning, Titus," she laughed. "You love me. Admit it." Paloma playfully pushed him. "Say it," she coyly demanded. With a smile, she turned and brushed her nose against his. "It's no secret," Paloma told him. "I already know." Then, she smiled at him and laughed. "You're such a man sometimes!"

What she said was true. Mr. Paraclitis was unsure how to express himself. Never had he experienced love. He didn't know how to respond.

As had other smitten Stellarii, Paraclitis visited Razzouk's tattoo parlor for an answer. Before evening, Paloma's name was inked upon his chest beneath a flying dove. The graceful bird flew across his heart with two red gerbera daisies in its bill. Paraclitis expected Paloma would understand his feelings for her when she saw his new tat.

That evening, they dined together at the village cantina. Alone in a darkened corner, he took her hands in his. The graphene dove encouraged him. "I love you, Paloma," he was going to tell her.

She leaned forward and pressed her lips against his. "Don't ever leave me, Titus," she adjured. Before he could reply, two harried Marine corporals burst into the establishment. "Lieutenant, you're needed immediately aboard the *Lioness*. Commerce raiders have attacked three convoys. Several ships have been lost. All Hades is breaking loose."

He rose and took Paloma in his arms. "Wait for me," he told her. Then, Paraclitis promised. "I'll be back soon." It was the promise servicemen rarely keep.

Seven weeks later, he returned. Paloma didn't greet him. Nowhere was she found. When he asked her friends, Paraclitis was told that Paloma passed recently from a sudden outbreak of emortualis virus. He received her personal effects from the medical center. Among them was a handwritten note. It read, "I always knew you loved me, silly." Smudged by lipstick, it revealed her kiss.

Never did Paloma see the tattoo which forever expressed Paraclitis 's love for her.

Now, eighteen months later, warships raced past the *Lioness* towards Semotus. Interstellar clouds of war gathered. Duty again called Mr. Paraclitis. This time, no one awaited his return.

The Lieutenant Commander mustered the ship's department heads to the ready room. The first to arrive was Lieutenant Condictor Sarcio. He was the ship's Chief Engineer. Sarcio was also a veteran of the deadly Barbary Wars. Throughout the fleet, he was known as "Greyhound," because he and his men had transformed the *Lioness* into the fleet's fastest ship.

Lieutenant Ferox Arma followed closely behind. He was the ship's Weapons Officer. Arma and Sarcio worked together to increase the power of the *Lioness's* pulse cannons to nearly twice that of other destroyers.

Behind them, followed Lieutenant Compertus Hippolos. He was Chief Navigator of the *Lioness.* Many thought Hippolos was the Navy's best navigator.

As Ensign Sonos approached, she was brushed aside by the ship's physician, Dr. Nixus Varus. Unlike the ship's other officers, Varus was not a veteran of the Barbary Wars, nor had he ever served in combat. He irately rushed into the ready room.

As Commander Clades entered, Miss Sonos updated him. "Commander, another forty-six ships passed us. They're following the same course." Clades wasn't surprised.

"Thank you, Ensign."

Mr. Clades addressed the Department Heads. "I called you here because I may order this ship into combat. Before I commit the *Lioness* to battle, I need your counsel. Lieutenant Arma, can you identify those vessels?"

"They look much like *Immitus* class destroyers."

"Mr. Arma, no such vessel has ever been built. The *Immitus* is merely a concept."

"Aye, sir. However, each of them appears to carry six interfector-type pulse cannons. Judging from emissions, they're hyper-charged with several gigawatts of power. I haven't seen anything like them, except on design boards."

"Then they're definitely warships?"

"It appears so. Most are destroyers. Some are battle cruisers."

"Commander," interrupted Lieutenant Sonos, "another six warships passed us. They're following the others."

Mr. Arma added, "Five of them are battle cruisers. The big one is an interceptor carrier. From its size and shape, I estimate it carries thirty-eight interceptors."

"Are there any other ships approaching, Miss Sonos?"

"No, sir. It doesn't appear so."

Mr. Hippolos warned, “Commander, they’re headed to Outpost Semotus.”

“Did they detect us?”

“I don’t believe so, sir. The *Lioness* was shielded by a planetoid’s reflective atmosphere.”

“Mr. Hippolos, set a pursuit course for those vessels. Keep the brightest sun to our back. That should mask our location.”

“Aye, sir.”

“Miss Sonos, notify fleet headquarters that sixty-four unidentified warships are approaching Outpost Semotus at flank speed.”

“Aye, sir. Command has been notified.”

Doctor Varus was visibly upset. He angrily leaned across the ready table and interrupted. “What are you doing, Commander? Our mission is to chart deep space. It isn’t to hunt alien spacecraft.”

“I appreciate your concern, Doctor. As the ship’s physician, I expect your counsel to avoid danger. However, sixty-four unauthorized warships passed us on a direct course to our planet's only deep space outpost. That's not likely a coincidence. It appears danger has found us."

Varus replied, "At least try to hail them, Commander."

"Very well, Doctor. Ms. Sonos, please hail the unidentified spacecraft again."

The Ensign hailed the warships four more times without reply.

"They're not responding, Commander."

The Doctor protested, "They may not have received our hails. Maybe they didn't understand them? We can't know. You can't assume those vessels are hostile."

"With due respect, Dr. Varus," Clades replied. "We must assume it."

"That's madness, Commander. If you react aggressively, it will certainly provoke a galactic war!"

"Doctor, Outpost Semotus is in danger."

"There's no reason to assume those ships intend to strike Semotus. It has no military value. The place is nearly uninhabitable. At most, a few thousand people live there. Every gasp of their miserable biosphere comes from obsolete, worn-out generators. They wouldn't have daytime without solar satellites. Semotus is nothing but a barren, shadowy asteroid in the middle of nowhere. It has nothing that anyone would want!"

"Maybe, the aliens disagree with you, Doctor? Surely, they didn't send an armada this far to visit a well-lit rock?"

Ensign Sonos spoke, "Commander Clades, I disagree with Dr. Varus. We must protect the Outpost. The Navy needs it for deep space communications. Without Semotus, Fleet Command won't be able to organize a defense of Gemina."

"Mr. Clades, listen to me," warned Dr. Varus. "You can't trust Miss Sonos. She has loved ones on Semotus. Her sister and aunt live there. They're the only family she has. The Ensign would do anything to protect them. You can't believe a word she says."

"Is that true, Miss Sonos?"

"What Doctor Varus says is true, sir. My sister, Tatiana, lives on Semotus with my aunt, Laelia. My father was killed aboard the *Cilicia* when Convoy PQ-17 was attacked by baryon marauders. When we learned his fate, Laelia took us into her home. Our mom died when the emortualis epidemic hit the outpost. They're the only family I have."

"I'm sorry, Lieutenant."

"Sir," she added, "My aunt teaches third grade. Next week is my sister's sixteenth birthday." Sonos implored the Commander. "We can't let those things attack Semotus. No one will survive. Please, sir."

"If those ships do attack the outpost," the Doctor protested, "there is nothing we can do to stop them. We are one warship against sixty-four. All of them are of equal or greater class than the *Lioness*."

The remark irked his fellow officers. To them, the *Lioness* had no equal.

Clades queried the Doctor. “So, you propose we do nothing?”

“No, Commander, we have a mission. We were ordered to chart space. That’s what we should do.”

Displeased with Varus’ comments, Mr. Arma interjected. “Commander, I agree with Ensign Sonos and disagree with the Doctor. Should the outpost be lost, our Navy won’t have adequate time to crew our ships. Half the fleet is orbiting Gemina with skeleton crews. Their senior officers are either on Rest and Recreation or they’re at our Military Operations Stadium. The Navy needs at least an hour to ready our ships for combat. Unless we delay the invaders here and now, those things will sweep through the sector unopposed.”

Dr. Varus loudly interrupted, “You assume they are hostile and call them ‘invaders.’ Answer me. Have the strangers powered their weapons?”

“No, Doctor.”

Varus continued, “Have they initiated defenses or tried to evade us?”

"No, sir."

"Doesn't that prove their intent is not hostile?"

Paraclitis answered, "Doctor, it only proves they've chosen not to reveal their intentions, capabilities, or weaknesses."

The Doctor shook his head in frustration. "Mr. Paraclitis, I understand Ensign Sonos' concerns, but you have no one waiting for you on Semotus. You have only a memory!"

"That's enough, Doctor!" Clades reproved.

Properly rebuked, the doctor replied, "Please accept my apology, Mr. Paraclitis. Commander Clades, forgive my ill-mannered remarks, but you asked for my counsel. I tell you, sir. Don't commit this ship and her crew to certain death."

"Mr. Paraclitis," asked Commander Clades, "What say you?"

"Commander, I don't seek armed conflict. None of us do. However, I don't think we have a choice. We must engage the invaders. Unless we delay them, most

Geminum warships will fight with undermanned crews led by junior officers. They won't have a chance."

"That's insane," argued Doctor Varus. "The beings in those vessels have shown no aggression. Surely, they didn't travel a gazillion light years to attack an asteroid. That's ridiculous!"

"Commander," Mr. Paraclitis replied, "it's highly unlikely that anyone would send sixty-four warships, including an interceptor carrier, on a reconnaissance-in-force mission. They're not sightseeing. That fleet is going to attack Semotus. There's no doubt about it."

Mr. Sarcio tried to enlighten Varus. "Doctor, we port at the Outpost twice yearly. Everyone has friends and loved ones there. I know every shop and restaurant as if it was my hometown. Whenever we dock, Chief Nux himself greets me and my engineers with the Sector's best, cold beer."

Dr. Varus grew more exasperated. "Lieutenant Sarcio would sacrifice us for free beer. Who can trust his judgment?"

Mr. Hippolos added, “Doctor, you don’t understand how valuable Semotus is to scientific discovery. Explorers gather there. They tell us about uncharted sectors they’ve visited. One of my academy classmates is assigned there to search for Gemina’s ‘Twin Planet.”

“Then, Semotus is a place for fools,” complained Varus. “The Twin Planet is a myth and sacrilege. Everyone knows that.”

Lieutenant Arma interrupted. “Commander, I have buddies on the Outpost. Many are gunny officers. They’ve tried to procure serious weapons, but all they got from Fleet Command were pea shooters. The outpost is defenseless against those warships. We’ve got to intervene. The *Lioness* is their only hope.”

Mr. Paraclitis addressed Varus. “Doctor, you’re new to this ship. The people of Semotus are not strangers. We won’t abandon them.”

Varus grew red with frustration. “Commander, there are ten billion people on Gemina who depend on us not to start a war. If one erupts, they will need this ship and every other in our Navy to defend them. Those ‘invaders,’ as you call them, outnumber our fleet fourfold.”

"Doctor," Clades answered, "The first duty of this ship is to protect Gemina. We will do so here and now."

"Don't be a fool, Commander. Retirement awaits you. Your family expects your return. Take this ship home, sir. Let your distinguished career end in honor."

"I'm not yet retired, Doctor. I still have duties of command."

"Think about your wife and daughters," Varus urged. "They want you home. That's true for us all. We have loved ones waiting planet-side."

"Who is waiting for you, Doctor?" asked the Commander.

Varus hesitated. He slowly sat down and rested his forehead in the palms of his hands. Quietly, the Doctor replied. "I envy you, Commander Clades. You have a loving wife planet-side. Would you deny her the comfort of your embrace, your touch? Your children expect you back, sir. They're secure in your love and eager to share their newest experiences with you. Surely, you hunger for the sounds of their voices calling?"

The Doctor sorrowfully continued. "I have none of that. My wife sued me for divorce. She was awarded custody of our children. I became a visitor in their lives. For months, I was angry and depressed. Then, I realized the Judge made the right decision. My professional obligations didn't allow time to raise a family. I missed my children's school activities, teacher conferences, dance lessons, even their birthdays.'

"That's why I joined the Navy. They promised to assign me to our hospital ship, *GSS Clementia,* where doctors are granted blocks of shore leave. I could spend time with my son and daughter. However, an Admiral's daughter-in-law took that spot from me. I was assigned instead to the *Lioness*. I wasn't supposed to be here. Mr. Clades, I haven't seen my children in months. All I wanted was to be part of their lives again, a better part."

"I'm sorry, Doctor."

"Commander Clades, please don't let me die as a stranger to my children. Please get me home. Get us all home."

"Doctor," answered the Commander, "The *Lioness* is the only Geminum warship in this Sector. There is no one else. Our children and loved ones depend on what we do here, today. We must give the rest of our Navy adequate time to crew our warships. Fleet Command must know the enemy's strengths and weaknesses. We can provide both."

"May the gods have mercy on us all," moaned Varus. The doctor slumped into his chair and whimpered, "We're all going to die."

Commander Clades continued his briefing. "Doctor Varus is correct. The *Lioness* is but one ship. Not even the mighty *Vindictus* could defeat those ships alone. However, we will engage the strangers.'

The Commander continued, "This is our plan. They don't know how many we are. Ms. Sonos, I want you to echo our previously recorded communications from every asteroid and space rock in the Sector. That will lead the enemy to believe we are many in number."

"Aye, sir."

"Mr. Arma, we will first strafe their middle column from behind. Target their propulsion systems. Hopefully, that will provoke a response."

Varus lifted his head and asked incredulously, "You want those things to attack us?"

"Doctor, it's the only way to determine their limits and capabilities."

Clades continued his battle plan, "Once we finish the strafing run, we will deploy mines at the head of their formation. That should damage or destroy more ships in their main body, as well as whatever ships pursue us. Our attack may diminish their strength by one-third. It may dissuade them from advancing further into our Sectors."

All the officers nodded in agreement, except for Dr. Varus.

Clades continued, "If the invaders don't withdraw, we will attack again. Mr. Hippolos, chart a course to the brightest sun. Its electromagnetic radiation will blind them. They will have difficulty separating us from any dominant solar object in the background."

"You can't know that," the doctor challenged.

"Of course, not, Doctor. We can only anticipate it." Clades further instructed Mr. Hippolos. "If they pursue us, put the *Lioness* in orbit around the sixth planet."

"Aye, sir," replied Mr. Arma. "They won't see us behind its particalized atmosphere."

"Once they pass," continued the Commander, "we will strike again, undetected."

"You can't know that, either!" Protested Varus.

"Doctor, we can't know anything without provoking a response. Only then can we inform our Navy about the enemy vessels' speed, weaponry, defenses, and abilities."

Varus protested. "You must be mad! Do you propose our deaths will warn others of their peril? Will we be the proverbial 'canary in the coal mine?'"

Miss Sonos answered, "We're no canary, Doctor. We're a *Lioness* and we have cubs to protect. That includes yours."

The Commander continued his battle plan. "We will then slingshot from orbit and strafe their port column. Mr.

Arma, we don't need to destroy them. We need only disable or slow them down. Target their propulsion systems. Our second attack may reduce their fighting strength by another third."

"Aye, sir," acknowledged Paraclitis. "With their strength substantially degraded, the invaders will surely withdraw."

The doctor was appalled. "I don't believe this crew will obey such reckless orders!"

Ms. Sonos replied, "As Communications Officer, I assure you that this crew is ready and willing to fight."

"No," groaned Varus, "every one of you is insane!"

Clades queried his officers. "Mr. Sarcio, is engineering ready?"

"Aye, sir. The ship is ready to prowl, pursue, and pounce on the enemy. We can run, if necessary. They will be hard pressed to catch us. There's no warship in the fleet faster than the *Lioness*."

"Is damage control ready?"

"Aye, sir. We're ready to take their best shot."

"Weapons control, are you ready, Mr. Arma?"

"Aye, Commander. We will give them our best shot."

"Is navigation ready, Mr. Hippolos?"

"Aye, sir. All courses have been plotted."

"Communications, please notify Fleet Command that we will engage the enemy fleet approaching Outpost Semotus. Tell them to stand by and receive our system readings. They will provide the enemy's strengths and weaknesses. It will give our Navy an opportunity to prepare a defense, if necessary.

"Aye, sir. Fleet Command, this is the *Lioness*. We are prepared to engage the enemy. We will engage them in sixty seconds. Stand by for system transmissions."

"Advise the crew, Ms. Sonos."

"Battle stations, battle stations. This is not a drill. We will engage the enemy fleet in 60 seconds: 59, 58, 57..."

"Mr. Arma, power all weapons."

"Aye, sir. All six pulse cannons are ready for action. Commander, each has six gigawatts of hurt waiting for those things. If they have any sense, they'll run from us as fast as they can."

"Dr. Varus, is med bay ready?"

"Are you asking whether we're ready to treat horribly wounded and dying crew members? No, sir, but we will do whatever we can, damn you!"

"18, 17, 16, 15..."

"Commander Clades," interrupted Ms. Sonos. "We've received a communication from Fleet Command."

"Let's hear it."

The voice of Prime Minister Morus boomed over the intercom. "This is the Prime Minister. Listen very carefully. I order the *Lioness* and her crew to stand down. You are not to engage the visitors. I repeat, you will not engage the visitors. That is an order. I am directing the

Lioness to rendezvous with Battle Cruisers *Essex* and *Renown* in Yellow Sector 14.[3]"

"Visitors?' Prime Minister, did you call them 'visitors'? With all due respect, visitors don't bring artillery to your doorstep. Those warships are hostile. If we don't engage them here and now, they will attack Outpost Semotus. There will be a great loss of life. The fleet will lose all deep space communications. Gemina will be in peril."

"Commander Clades, let me make myself clear. We will not begin a galactic war. The visitors have not attacked us, nor have we provoked them. I am ordering you to stand down. Rendezvous with Admiral Forctis in Yellow Sector 14. That's an Order. Did I make myself clear?"

Reluctantly, Clades answered, "Aye, sir."

Ensign Sonos reported, "The Prime Minister's communication has ended, Commander."

---

[3] Sectors are identified by color. Outpost Semotus is in an occupied Sector, so it is a Green Sector (16). Sector 14 is unexplored, so it is Yellow. A Red Sector is one that is too hazardous to enter.

"Thank the gods!" Doctor Varus exclaimed. "Finally, the voice of reason has spoken."

"Commander," Miss Sonos pleaded, "we're not going to withdraw, are we? Please, sir, we can't abandon Semotus. They're counting on us. We're their only hope!"

Lieutenant Arma added, "The crew is ready to engage the enemy, Commander." Other officers agreed.

Mr. Sarcio cautioned, "We can't let those things have a free shot at Gemina. That will happen if they knock out Semotus."

Clades sank deep into his Captain's chair. He believed the Prime Minister's order was unwise, even foolish. With bowed head, he told his crew. "We have orders. It is our duty to follow them. Mr. Hippolos, chart a course to Yellow Sector 14. We will rendezvous there with Admiral Forctis and the *Renown*."

"But, sir," implored Miss Sonos, "the enemy fleet will reach Semotus in hours. There's no way they can defend themselves. Everyone will die. Please, sir, let us fight!"

Clades sadly shook his head in disgust. An order had been issued. As a Naval Commander, it was his duty to follow it and trust his leaders' wisdom. Clades wished he had never informed Fleet Command of his decision to engage the "visitors." It was a mistake to allow inept, misguided civilians to overrule his trained military judgment. He would never again make that error.

The *Lioness* and her crew left Green Sector 16, as directed. They grouped with the *Renown, Essex*, and other Geminum warships in Yellow Sector 14. Outpost Semotus was undefended.

[End of Chapter]

CHAPTER TWO

Attack on Outpost Semotus

A cold, lonely death neared Lieutenant Rimor. He piloted a Mark Nine Cutter through deep space. His small reconnaissance vessel suffered catastrophic quantum circuitry failures. Critical systems had shut down. Others were failing. Among them was life support. Without it, Rimor would have no oxygen to breathe. At temperatures falling to -270 degrees Celsius, he would quickly suffer severe hypothermia. It wouldn't kill him. In a ship without air pressure, his internal organs and skin would expand. Once that process ended, they would freeze, then totally evaporate. He would die in less than a minute.

He transmitted emergency broadcasts from his comm system. "This is Lieutenant Rimor." There was no answer. He set his transponder's distress signal and hoped it would be picked up by space tracking devices. He didn't know where he was, nor whether anyone was within his transmission range. "This is Lieutenant Rimor," he repeated. "I'm piloting the Geminum Cutter, *Explorator*. I'm experiencing navigation and communication system failures. All systems are shutting down. I'm requesting assistance. Repeat. I'm requesting assistance."

There was no response. Rimor launched his navigation buoy. It was unlikely to provide accurate coordinates. However, its signals might be received by Outpost Semotus. That was the only base within days of his last known position.

"Outpost Semotus, this is Lieutenant Rimor aboard *Explorator*. Do you read me?"

His query was met with more silence. Lost in space, his options quickly dwindled. Rimor had made astonishing discoveries on his mission. They were world changing. However, all would be in vain if he wasn't soon located.

In the dark loneliness of space, the Lieutenant contemplated his fate. Rimor began to record his last message. "Gemma, this is Rimor. There's something you need to know..."

His recording was interrupted. The Cutter's communication module crackled. "*Explorator*, this is Outpost Semotus. Do you read?"

Rimor received the reply with jubilation. He answered, "This is Ensign Rimor. I read you. Can you give me your coordinates?"

They didn't receive his transmission. Again, his query went unanswered. He anxiously awaited another message.

"*Explorator*, this is Outpost Semotus. Do you read?"

They couldn't hear him. He tried again to broadcast an answer without success. After a moment of space noise, his comm crackled a message. "*Explorator*, this is Outpost Semotus. We have you on QDRS[4]. If you read us, follow the coordinates we are transmitting."

To the Lieutenant's surprise, he received the information. The Outpost was within an hour's travel. His ship could complete the journey on emergency power. He would survive.

---

[4] QDRS is an acronym for "Quantum Detection and Ranging Systems."

The Lieutenant had made astonishing discoveries. Gemina would celebrate them and him. Yet, a greater discovery awaited Rimor on Semotus across a lonely table in a cafe. It would forever change his life.

Rimor was assigned to Outpost Semotus. It had been intended to be a science outpost. However, as more ships ventured to deep space, the Outpost's hangars, landing pads, and repair facilities drew galactic vessels of every kind. It became a small but busy space port. The Lieutenant's cutter was frequently serviced there.

The outpost was a craggy, rocky asteroid in the middle of nowhere. The place's machinery and equipment were maintained by skilled, but overworked servicemen. Its artificial atmosphere was generated by obsolete, worn machinery.

Behind its base stood a tall, rocky mountain. The towering crag's long shadows allowed the illusion of night to an asteroid that was constantly illuminated by three orbiting satellites. It also offered a pretense of safety.

The Lieutenant carefully glided his Cutter to port facilities. He was eager to land. The place had a population of twenty thousand people, or so. Few knew him, but he expected a special welcome.

Lieutenant Rimor wasn't just a pilot. He was also an archaeologist. He was assigned to search deep space for a mythical planet. Ancient legends foretold that his world, Gemina, had a twin. The name "Gemina" meant one of two twins. The notion of a twin planet was heresy to some. They feared it was a deification of Romulus and Remus, the twin brothers who founded Rome. Others ridiculed it as a foolish notion. Few believed a twin planet existed.

There were rumors among his classmates that Rimor's assignment was a cruel joke. He was unpopular among Academy graduates for wasting his naval education on an archeology degree. They punished him by sending him on a fool's mission far from Gemina.

Eagerly awaiting Lieutenant Rimor was Ensign Gemma. She was an attractive, young Aviation Inspector. The Ensign hid herself in work clothes and protective apparel. Few saw Gemma without her hearing-protectors, eye gear, and baggy coveralls. She wore her raven black hair short, above her shoulders, so she could tuck it unseen in her helmet. Gemma didn't want to be noticed by anyone

except Lieutenant Rimor. Yet, no disguise could hide her blushing cheeks or bright smile. Eventually, she would be discovered. People were searching for her.

Gemma first met Rimor months before when he landed *Explorator* for maintenance. Initially, the young Lieutenant didn't impress her. She assumed he was another arrogant pilot. Gemma hurried past Rimor in the South hangar. For some reason, he called to her, "Hey, you're beautiful!" Her first reaction was to ignore Rimor and keep walking. Gemma assumed he was another jerk. However, as Gemma told it later, she "stopped to listen anyway. Because, you got to admit, the man had a valid point." The two quickly became inseparable.

Gemma couldn't hide her beauty from Rimor. He saw beyond her work gear. Her sparkling green eyes and quick smile always found him. He found beauty within her, too.

Gemma gazed deeply into Rimor's heart. She knew they were in love.

During the Lieutenant's many stops at Semotus, he and Gemma shared every moment. She once daydreamed that he landed his Mark Nine Cutter in her bay. As he disembarked his ship, she ran to him. Rimor removed his helmet and tossed it aside. With a radiant smile, the handsome young Lieutenant wrapped his arms around her waist. He held her off the ground and kissed her passionately. It was a crazy fantasy. She hoped it could be true. She wished it were true.

Communications announced *Explorator's* arrival. Over crackling transmissions, Gemma heard Traffic Control grant Rimor permission to land. She watched Rimor's cutter glide to the West Launchway. Ordinarily fast and graceful, the Cutter was a bit wobbly and unresponsive. That was uncharacteristic of a lightly armed reconnaissance ship. However, the sight warmed her heart.

Among other things, Rimor was a skilled pilot. Although his ship was compromised, he landed *Explorator* without incident. Gemma rushed to the Bay. Her cheeks blushed as her heart fluttered. He was the only man she had ever loved.

Chief Petty Officer Nux approached. He stood by her side as she awaited Lieutenant Rimor. He disliked the young pilot.

The Chief was the group's head mechanic. Nux was a big, strong non-com. He had calloused hands and scuffed knuckles. His coveralls were stained by grease and hydraulic fluids. Colorful tattoos danced along his arms whenever his bulging muscles turned wrenches or lifted tools. There was no one tougher than Nux. No one loved Gemma more than him.

Rimor disembarked *Explorator*. The Chief shook his head in disgust. He wiped a smear from his chin and spat on the ground. Nux grumbled, "Here comes that idiot, Rimor, again."

Gemma adored the Chief. When they first met, she was a lowly teenage Stellari. Nux accepted her on his crew, although she had no skills. He patiently taught her to be a mechanic. Eventually, Gemma's veins pulsed with mechanic's blood. She used safety wire as a toothpick, like the rest of the crew. When asked about "R&R," most understood it to mean "Rest and Recreation." However, as did other mechanics, Gemma assumed it meant, "Remove and Rotate." Rude pilots called her a "grease monkey." Others said she was a "tin bender." However, Gemma liked those terms. She rose in rank to Ensign. Everything she became was owed to Chief Nux.

Naturally, the Chief's angry words about Rimor upset her. Nux was the father she never had. She loved both men and expected Chief Nux and Rimor to be friends. Gemma tried to temper the Chief's remarks by appealing to his sense of duty. "You know Chief," she grinned, "you shouldn't talk that way about a superior officer."

"Gemma," he replied. "The guy's a joke. He's an embarrassment. He's a loser. Command sent him out here because nobody likes him. You should forget him."

"Chief, they sent *us* out here, too! Are we losers?"

"Come on, Gemma. The jerk's an archeologist! There ain't no archeology out here. There's nothing. That's why they call it 'space!' He should be planet-side digging in dirt. Maybe, he could be chasing Helen of Troy, or something?"

"So, you think he's here chasing me instead?" chided Gemma.

Chief Nux cautioned her. “He ought to be chasing his shadow. We don’t need him here. You should stay away from the guy. He’s no good for you.”

Gemma gave Nux a cross look. “Chief, why are you so mad? You’re acting like Rimor messed with your toolbox!” To a mechanic, there’s nothing worse.

“I’m tired of looking at him,” answered Nux. “He prances around like he’s some great explorer. He even calls his ship, *Explorator*.’ I wish he would discover something and go back to Gemina.”

“You know he’s searching for the Twin Planet. Rimor told me he thinks he’s getting close.”

“There ain’t no twin planet, Gemma. It’s a fairy tale. He’s an idiot. Everybody knows that. If it is real, he needs to find the place and go away. I’m tired of seeing his ugly face.”

“If he finds the twin planet,” answered Gemma, “Rimor will be heralded as a great explorer. He will be rich and famous! They may even name the new world after him.”

"Naaah, they ain't naming no planet *'Stupid.'*"

"I'm serious, Chief. Lieutenant Rimor is on the verge of being a historical figure!"

"Listen, Gemma. He ain't no astronomer. He ain't a scientist. He's hardly a pilot. If you ask the guy to fly, he just goes *bzzzzz*. He's an idiot."

Gemma grabbed the Chief's well-muscled arm. With raised eyebrows, she asked him, "Look, Chief, you're my best friend on this rock. Please cut Rimor some slack. Don't do it for him. Do it for me, okay?"

"Gemma, I know you're sweet on the guy. I don't like it, but I'll give him a break. I won't use his head to shine my boots. I won't promise nothing else." The Chief picked up his wrench and left for the mission room. He expected to find Rimor's "*gripe sheet*" there. That's what they called pilots' reports of suspected mechanical issues. Nux particularly relished Rimor's gripe sheets. It gave him opportunities to humiliate the young Lieutenant. He expected his sarcastic replies would convince Rimor that he wasn't welcome. When answering Rimor's gripe sheet, the Chief answered:

Pilot: "Distance Measuring Equipment volume unbelievably loud."

Nux answered, "DME volume set to a more believable level.

Pilot: "Aircraft handles funny."

The Chief replied, "Aircraft warned to straighten up, fly right, and be serious."

However, Nux's favorite gripe was Rimor's remark about a misfiring engine.

Pilot: "Propulsion Engine is missing."

Nux answered, "Propulsion Engine found in engine pod after brief search."

Unknown to Chief Nux, Rimor enjoyed his solutions. The Lieutenant thought they were funny. He once reported a "stuttering engine," because he wanted to read the Chief's response. He knew Nux's reply would be hysterically funny, and it was. It brought a grin to the Lieutenant's face. He thought he and the Chief shared a sense of humor.

Rimor also appreciated the Chief's work. He and his crew kept *Explorator* ready and able for the longest journeys. Their work was so good that the Lieutenant once suspected they wanted him to go as far away as possible. However, the Chief's humorous replies to his gripe sheets convinced Rimor that the crew enjoyed his stays.

"There he is," snarled Nux. *Explorator* was parked in the West Hangar. Gemma anxiously awaited Rimor. As soon as he deplaned, she waved at him and called, "Welcome back, Lieutenant. We're glad to see you again." More accurately, *she* was happy to see him.

In disgust, Nux turned and left. He had better things to do.

Rimor heard Gemma and turned. His face glowed when he saw her. The young Lieutenant held open his arms. She had never seen him as happy. Gemma wondered if her daydream was coming true! "Gemma," he called to her, "I have something important to tell you. I need to tell you now. Can we go somewhere and speak in private?"

Gemma's heart raced. She knew they loved each other. From the first time Rimor came to Semotus, they spent every moment together. They shared their thoughts and dreams. Gemma knew Rimor was finally going to confess his love for her. Afterwards, she would confess hers. Never had Gemma been so excited.

Hand in hand, they walked to the Cantina. The two were having a wonderful time. Gemma was laughing and smiling. She knew Nux and the guys would ridicule her silliness, but she didn't care. Suddenly, the Outpost's sprinklers erupted. It was the closest thing to weather happening on Semotus. The couple strolled through what felt like buckets of rain. Initially, it surprised them, but the two continued without pause. At a busy intersection, they stopped for traffic to pass. Each looked deeply into the other's eyes, both dripping wet. Rimor put Gemma's face in his hands and kissed her passionately. She knew it was love. He was going to admit his love to her. She knew it.

The couple entered the cantina soaking wet. A hostess handed them towels to dry themselves. Rimor searched for a private space. He took Gemma to a table in a dark corner. She beamed with happiness. Rimor had never shown such joy. Gemma expected they would soon share an unforgettable moment. The dashing young Lieutenant held Gemma's hands in his. Rimor searched deep into her eyes for the beauty that mattered most to him. He happily smiled. Each was ecstatic. Gemma yearned to hear Rimor profess his love. "Gemma, I love you," he would say.

Rimor cleared his throat. He held her hands tighter and looked around the room carefully. When he knew no one was listening to their conversation, Rimor leaned towards Gemma and spoke quietly in a hushed voice.

"I found it, Gemma."

Those weren't the words she expected. Was he trying to say he found his love for her?"

"It's there. I found it. I finally found it."

Gemma was confused. What was he saying? He was supposed to say, "I love you!"

"Gemma, I found the Twin Planet. I did it!"

She was hurt, disappointed. Gemma yearned for words of love. She wanted romance, not astronomy.

"Gemma, I found the Twin Planet. It exists. It *really* exists."

"I, I really don't know what to say," she stuttered.

"It's identical to Gemina in size, mass, climate, and atmosphere. It's populated by people like us. They *look* like us. They even speak our languages! Can you believe it?"

Gemma was stunned. She could hardly speak. Rimor held her hands in joy while talking about a distant planet! "I don't believe what I'm hearing," she replied. Not only was she denied his proclamation of love, but Gemma realized that Rimor's discovery would take him from her. She expected Rimor would return to Gemina, where he would be heralded for his astounding discovery. She would never again see him.

Rimor didn't realize Gemma was upset. He thought her eyes watered with happiness for his historic achievement. Rimor gently took her hands in his to celebrate.

Gemma was disappointed and hurt. She assumed he would return to Gemina without her. Yet, their moment really was intimate. Rimor had shared his astounding discovery with no one else, not even Fleet Command. Gemma was the only person he told. More than anyone, he wanted to share his success with her. Rimor expected Gemma would be excited and happy. He soon learned otherwise.

Gemma complained, “Lieutenant, I don’t understand. You explored an endless vacuum and discovered an entire world. Yet here we sit. You hold my hands, but you can’t find what is before your eyes? Why don’t you see me, sir? Has it meant nothing to you that I met your cutter whenever you landed? Whenever you returned, I wore uncolored safety glasses, so you could see the eyes that love you? Didn’t you notice me? My work requires me to wear boots, gloves, and protective gear, but when you came, I wore my fluorescent red workwear and reflective yellow trim, so you would see me! How could you not notice? Am I too plain for your touch?”

Rimor was startled by Gemma’s words. He didn’t know how to respond.

Upset, Ensign Gemma rose. She turned so that Rimor wouldn’t see her wipe away tears. Without further discussion, Gemma left. The Lieutenant sat alone. He wanted to make things right. He didn’t know how.

Gemma hurried outside. The ground was still wet. She needed to be alone. There was one place where she most expected solace – her hangar. The Chief and crew wouldn't be there at such a late hour. No one would. She would share her heartbreak with the hangar's lonely darkness. The place welcomed her with silence and emptiness. It mirrored her feelings. She sat alone on a lift.

Starlight creeped inside. Something glimmered beneath Rimor's ship. She walked to the Cutter and found a wrench. Gemma recognized it. It was the Chief's wrench! She boarded the vessel. Its systems had been tampered with. Chief Nux had altered *Explorator's* navigational coordinates and records. Gemma was furious. Twice in one evening, her heart was broken. She exited Rimor's cutter almost fifteen minutes later.

Gemma sat alone in shadows, angry and sad. She wondered what to do. After much contemplation, she rose to her feet and marched to the chief's quarters. With clenched fist, she beat on the door.

"Chief, this is Ensign Gemma. Come forward."

A minute passed. Nux rose from bed and came to the door. He wore long johns and an old robe. "Aye, ma'am?"

"Chief, I know what you did. I found your wrench under Rimor's cutter. For the god's sakes, how could you do that?"

Nux stuttered.

"Chief Nux, don't you know you committed a felony?"

"I … I don't understand, Ma'am."

"Don't give me that 'ma'am' baloney. Why did you tamper with Rimor's navigation and communication systems? Why?"

"I'm sorry, Gemma. I'm really sorry."

"How could you do that?"

Chief Nux paused as he concocted a reasonable excuse. "I was just pranking the kid. Yeah, I was pranking him!" What Nux said was untrue. He had maliciously tampered with *Explorator*.

"That was more than a prank, Chief. You would have ruined the Lieutenant's career. You would have humiliated him. He would have returned to Gemina with altered data that made it appear that he mistook *Gemina* for the Twin Planet. They would brand him as a fool!"

"I didn't want to hurt you, Gemma. I just wanted him to go away. I wanted them to ground him, so he would never come back."

"Chief, that's horrible. It's *criminal*! I don't understand why you did it. You know how I feel about Rimor. Don't you know how I feel about you? Don't you know what you mean to me?"

Nux hung his head. He didn't want Gemma looking into his eyes. He was afraid she might see them watering. It was time for truth. "I ... we... don't want to lose you, Gemma. If the Lieutenant becomes a big hero, he's gonna ask you to go back to Gemina with him. He'll take you from us. You don't understand. We need you, Gemma. We need you *here*."

"Surely, I'm not so important?"

"I'm sorry, Gemma. I understand your feelings for me and the guys. They didn't have nothing to do with it, okay? Please forgive me. I'll fix it."

"I already fixed the systems." Gemma lifted Nux's big head by his chin. She asked, "Chief, did you notice Rimor didn't tell anyone else about the Twin Planet? He didn't notify Fleet Command, news organizations, no one! He discovered this incredible planet, and I'm the only person he told."

"I seen that, Gemma. You're important to him. I know he's important to you. What I done was foolish and wrong. Please forgive me."

"Chief, I love Rimor, but I can't return to Gemina with him or anyone else. You're stuck with me."

"Maybe Rimor can stay here? There's probably more planets out there to discover? I'll treat him right. I promise."

"Chief, I also love you and the guys. Thank you for caring about me. Rimor hasn't asked me to leave with him. He hasn't discussed his return to Gemina. I don't think I gave him a chance."

"Gemma, give the kid a break. I see how much you love that knucklehead. Talk to him."

The Ensign kissed the gruff Chief's cheek. "Good night, Nux." She returned to the cantina. Rimor was gone. Gemma rushed back to the hangar. *Explorator's* cockpit was lit. She knew Lieutenant Rimor was aboard. Gemma anxiously tapped on the hatch. There was no answer. She tapped again, more loudly. Slowly, the hatch opened. Rimor climbed down.

He apologized, "I'm sorry, Gemma. I didn't mean to hurt you. I would never hurt you."

She answered his apology by grabbing him. She pulled Rimor close and kissed his lips passionately. "I would never hurt you, either. I didn't understand, because I was listening for something else. Please forgive me."

"Gemma, I want you to return home to Gemina with me."

"I can't go, Rimor. I love you, but I can't go. Don't ask me why."

The Chief's voice interrupted them. "*You got to go, Gemma!*"

Gemma was surprised. She didn't expect the Chief. He was short of breath from hurrying to warn her. Puzzled by the urgency of his remark, she asked, "What do you mean, Chief?"

"You got to get off this rock. You got to go, now! The Shore Patrol is coming for you. I heard it on communications. Security is on its way to take you into custody."

Rimor didn't understand what was happening. "Chief, why is the Shore Patrol coming for Gemma? What's going on?"

"They say she's a fugitive. They say she's a convicted felon, and they're coming for her."

"That's crazy!"

"They said Gemma stole some big-shot Captain's DUK.[5] It's a felony. She was tried in a Navy Court. They sentenced her to ten years imprisonment."

"That's ridiculous. No one gets ten years in prison for stealing a DUK! I don't believe it."

"Lieutenant, it's worse than that. The ten-year prison sentence is supposed to be served at Ultimo Prison on Cancere Isle. It's a worse place than hades. Gemma would already be there if she hadn't escaped. She's a fugitive. They charged her with escaping, too. They'll put Gemma in prison forever."

"Gemma, is what Nux said true?"

"The Chief is right. I *am* a fugitive. When I was fifteen, I dated Captain Pullum's son. His name was Amator. He was a freshman at the Naval Academy. There were rumors that our relationship was sexual. That would have been statutory rape because of my age. Amator could

---

[5] A DUK was a military aerial vehicle used to transport troops and equipment. The acronym was given to it by servicemen. It stood for, "Drab Ugly Kite."

have been kicked out of the Academy and put in prison. His Dad wasn't going to let that happen.'

"One afternoon, Amator and I went to the Lake. He took his Dad's DUK. When we returned, there was Shore Patrol waiting. When I exited the DUK, they arrested me. I was charged with felony larceny of a military vehicle. Amator protested. He told them he piloted his Dad's DUK. He told them it wasn't stolen. They refused to hear him. He yelled. He tried to stop them. They ignored everything he said. They pushed him to the ground. I was taken to the brig and locked up. The Court appointed an attorney to defend me. She didn't listen to a word I said. She pled me guilty. They sentenced me to imprisonment of ten years in Ultimo Prison. That's it."

"How did you escape?"

"Amator paid off a guard. My cell was unlocked. I walked out of the brig through the front door. The guy sat at his desk and acted like he didn't see me. Now, I'm charged with escape, too. I'm probably looking at another ten years."

"But Amator can straighten it out? He's an officer, now."

"I'm sure he would, if he could. However, Amator is assigned to deep space. He escorts merchant convoys to protect them from raiders. Even if he could help, I wouldn't involve him. There's enough trouble already. Some idiot would charge him with aiding and abetting my escape!"

Rimor was stunned. It was difficult to absorb what he learned. "But how did you get into the Navy?"

"I used a fake ID. It's the same one I used to buy beer when I was a kid. It's easy to get in the Navy. Two doctors stand on each side of you. They look in your ears. If they can't see each other, you're a stellari."

"Is your name really Gemma?"

"That's my name, Cassia Gemma. I got assigned to Outpost Semotus to get as far away from Gemina as possible. I hoped no one would ever discover me here. Hey, I made it all the way to ensign before they found me."

"The Chief interrupted. "You've got to leave, Gemma. They'll be here in minutes."

Rimor took Gemma's hands in his. He pleaded, "Come with me. There's room in *Explorator*."

"I can't do that, Rimor. Honor and fame await you on Gemina. The world will celebrate you as a legendary explorer. They will welcome you as a hero, but not with me. I'm a convicted felon there. Once I disembark *Explorator*, military police will imprison me and cause you embarrassment. Just go. I love you, big goof. Go back for me."

"Gemma," Rimor urged, "you've got to come with me. You thought I couldn't see your love? I saw it when I first saw you. I looked forward to being here with you more than anything. That's why *Explorator* needed so much maintenance.'

"I remember the first time I ever saw you, Gemma. There was nothing wrong with my ship. You signaled *Explorator* to a hangar. As soon as I deplaned, you took off your helmet. The first thing I saw were your beautiful, olive-green eyes. Your bright smile welcomed me. As you walked away, I called out to you. I don't remember what I said, but you stopped and laughed at me. You replied, 'Welcome to Outpost Semotus, Mr. Explorer. You're not lost, are you?'

"You told me where to get hot coffee. I went there and found you sitting at a table with two cups of coffee. You shook your head. "You might as well have a seat, Lieutenant. Your cutter is a mess. You might be here for a while.' We've been together ever since.'

"Gemma, you told me that only the best mechanics are stationed at Semotus. You said that's why you're here. You said you were too good for anyone else. Everything you said was funny. I never had so much fun with anyone else. There's nothing I enjoyed more than returning to you. After weeks of deep space, you brought light to my darkness.

I fell in love with you when we first met. I loved you then. I love you now. You inspired me to find the Twin Planet. No one else believed in me. I found it for you."

"You fell in love with me?"

"I love you, Gemma."

"Why didn't you say it?"

"Gemma, I'm not the only one who loves you. Chief Nux loves you, too."

The Chief bowed his head.

Rimor spoke to Nux. “That’s true. Isn’t it, Chief?”

He nodded his head slowly. In a hushed voice, the Chief replied, “Yes, it’s true, Gemma. I love you.”

“*You* love me, too?”

“I love you, Gemma. Now, please leave with Rimor before the SPs get here.” Gemma was the daughter Nux never had.

Rimor pulled on her. “Gemma, come with me. I’ll take you to the Twin Planet. We can live there together.”

“I can’t allow you to forego fame and honor because of my misfortune.”

Suddenly, defense sirens sounded. First, one, then another. Eventually, there were six howling at the same time. Activity stirred everywhere. People sprinted from one place to another. Many looked at the sky.

An urgent voice boomed over loudspeakers. “This is an alert. This is *not* a drill. I repeat, this is not a drill. Take your battle station. All non-military personnel take shelter.

I repeat, all nonmilitary personnel take shelter *immediately*."

A vehicle roared past. The driver yelled, "They're coming! Get ready."

There were abrupt, bright flashes in the sky. Several merchant ships exploded as they frantically attempted to descend through the Outpost's atmosphere. One dived recklessly. Its nose glowed red from friction. Its hull was torn from weapons fire. The ship tried to land on the Northern Strip. An alien interceptor chased it. Automatic guns tried to ward the pursuer away from the wounded craft, but the attacker slipped past them. It obliterated the merchantman two hundred feet above the Outpost. Parts, pieces, and shards of metal crashed to the ground near Gemma. Debris bounced and spun past her.

Nux ran to an anti-aircraft gun. He loaded it and turned it on a swivel. One of his crew joined him. Another manned a second gun. When an interceptor buzzed the field, the Chief blasted at it. He hit it many times, without effect. Gemma ran to a third gun. She raised it to the sky and found a target. She unleashed a vicious burst of gunfire. Several shots found their mark, also without effect. The things were impervious to their weapons.

"Get out of here," yelled Chief Nux. "Gemma, get the hell out, *now*!"

Gemma continued firing her gun. She wouldn't leave Nux and her crewmates. Two interceptors swept low across the airfield. They strafed buildings, which burst apart in flames. It was a terrible sight. Men and women ran from burning barracks. Some were afire, screaming and flailing at the flames that engulfed them. Other guns hit the strafing enemy vessels, but the attackers were undamaged. Nothing seemed to harm them.

The Chief abandoned his weapon. He pulled Gemma from hers. "I'm begging you, Gemma. Please leave now! Our own people want to put you in prison. Those things want to kill you. Get the hell out of here!" He turned to Rimor and yelled, "Get her out of here, Lieutenant!"

One of the crewmen called out, "Hey, Gemma. I love you, too. Now get the hades away from here!"

The third mechanic hollered. "Hey, Gemma, we all love you. When they put you in prison on Gemina, say hello to my old lady. My in-laws had her busted for felony stupidity after she married me. I bet she's waiting for you in solitary! Now, get out of here!"

Nux repeated, "Go on, Gemma. Get the hell out, please!"

Rimor took her hand. He pulled her from the gun. She watched Nux and his crew blasting away at the attackers. Nux roared, "Go, *now*!"

Gemma and Rimor ran to his waiting cutter. As they approached the ship, a nearby hangar burst into flames. Ships on the launchway exploded one after another, as swooping interceptors unleashed their vicious weaponry. "Quick, get in!" yelled Rimor. Once Gemma entered the Cutter, the Lieutenant followed and grabbed the ship's yoke. He took its controls and awakened *Explorator's* powerful engines.

The ship screamed with rage as they raced from the hangar down the launchway. An enemy interceptor tried to strafe them, but Rimor jammed his breaks, causing it to miss and overshoot them. After skillfully evading the attacker, he accelerated his Cutter and lifted far above the airfield. Another interceptor tried to chase them from behind, but *Explorator* was too fast. She ascended quickly into space. As they fled into the expanse, Gemma watched Outpost Semotus being destroyed. "They're being murdered," she cried. "Why would anyone do that?" She was inconsolable. Everyone she loved there was dead or dying.

Two more enemy craft targeted *Explorator*. They fired several weapons' bursts, all of which fell short. The attackers lacked the cutter's speed. Quickly, they faded behind and abandoned pursuit.

The battle of Semotus ended tragically. Rimor raced away with Gemma. Whatever their speed, they couldn't escape memories of the Outpost's terrible destruction.

Yet, a new world awaited.

[End of Chapter]

CHAPTER THREE

Promises of Unwounded Blood

Gemma's world was shattered. Outpost Semotus was destroyed. Everyone she loved was killed in an alien attack. A journey across the vast expanse of space would take her and Rimor to a new world. It was nearly identical to Gemina. Rimor said some natives called it "Earth." Others said it was "Terra." Whatever it was called, their lives would begin anew there.

They traveled at speeds many times faster than light. However far they traveled, Gemma's broken heart couldn't escape Outpost Semotus. She was inconsolable. Tears of sadness streamed down her cheeks. Rimor wanted to brighten her mood, but it wasn't the time. She needed to grieve.

Shortly after they fled Semotus, *Explorator* received reports from the Fleet. They weren't encouraging. The monsters who attacked the outpost also destroyed six Gemina warships which had been sent to defend it. That included the Navy's two powerful battle cruisers. The remaining three vessels withdrew with heavy damage. The *GSS Lioness* fought their pursuers to a standoff during their retreat. Yet, the entire enemy fleet pursued them.

The attackers seemed invulnerable. Only nine Gemina warships remained between them and the home world. The planet's chances of survival weren't promising.

Gemma finally spoke. Her voice trembled. "I should have stayed with the Chief. I should have fought with my guys."

"Don't say that, Gemma. They loved you. You're the closest thing the Chief had to a daughter. The others loved you like a baby sister. They wanted you to live. They couldn't visit the afterlife without knowing you were safe. You did the right thing. You allowed them death with dignity."

"I miss the Chief and guys," Gemma told Rimor. "They were *my* knuckleheads." Gemma forced a broken laugh as if fond memories could return them to life. "I loved them all!"

After pausing to regain her composure, she continued. “But I have nothing from my guys to remember, except the Chief's wrench. I never gave it back to him. He will probably haunt me forever.”

“I doubt it. There is no one else whom the Chief would rather have his wrench than you.” Rimor took Gemma's hand. "Let's take the Chief's wrench to the new world. We can build new lives together.”

Gemma regretted that Rimor would never be honored for his discovery. She didn't want to cost him his future. “Let's go back, Rimor,” she urged.

“We can't. There's nothing on Semotus but death.”

“I know. Let's go back to Gemina.”

“Gemina couldn't have survived those things. If it did, only prison awaits you there.”

Rimor dropped *Explorator* from FTL. Gemma was surprised at the sight. She recognized the solar system. They passed Uranus, Neptune, Saturn, Jupiter, Mars, and approached Gemina?

“Rimor, we’re home. You brought us home?”

“No, Gemma, it may look like home, but it’s not.”

“How could that be?”

“I can’t say, but that’s not Gemina. It’s the Twin Planet.”

“No, that’s Gemina! We’re back.”

Rimor pointed across the view screen. “Look at the moon, Gemma.”

Gemma marveled at the sight. She was surprised. "What you say is true. Our moon hasn’t looked like that for sixty years. It’s amazing. They haven’t yet discovered her mysteries."

"No, they're far behind us in technology."

“What about the people there?”

“They’re not much different from us, Gemma.”

“How do they treat ‘visitors’ from other planets?"

"They’re unaware of visitors.’

"Rimor, you mean we'll be the first? Will they treat us like oddities? "

"No one will know we're from another world, Gemma."

Gemma nodded her understanding. She managed a curious smile. "Are you telling me that my fake ID will work there?"

Rimor laughed. At last, Gemma was finding her humor again. "I think your fake ID would work anywhere. However, I manufactured new IDs. We will have new identities and backgrounds when we arrive. They'll think we're natives."

"Will I really be able to understand their speech?"

"You'll understand most of it, Gemma."

"How are they different from us?"

"We're technologically superior. Maybe, we're more civilized, but their arts and music are advanced beyond anything we've imagined."

"Who cares about 'arts' and 'music?' They build nothing, solve nothing. They have no value."

As they spoke, the Lieutenant guided *Explorator* into a graceful descent. Terran defenses wouldn't detect it. The cutter's highly advanced technology was far advanced beyond anything they had.

The Cutter landed in a remote, wooded area outside of New York City. Rimor programmed *Explorator* to camouflage herself. He also hid the ship with tree limbs, brush, and dirt. Gemma stood guard with her pulse rifle as he finished. The shoulder strap of her canvas satchel nestled tightly between her bosoms.

They didn't expect trouble, but Rimor was also armed. He carried a pistol, a battle knife, and a pulse rifle. The two began their first Terran journey. They couldn't anticipate how it would end.

Gemma needed to tell Rimor she loved him. However, her heart was overburdened with sadness. She hoped he would understand. "Please be patient with me, Rimor. You have my heart, but it's broken. Let me heal."

"I grieve, too, Gemma. I know it sounds crazy, but I always thought the Chief liked me! I didn't know he thought I was a cad who wasn't good enough for you!"

Gemma answered, "Nux didn't think you were a cad. He thought you were a knucklehead." She managed a broken smile at the Chief's memory.

Rimor was familiar with the path. He used it after his first landing on Terra. It led to a large shopping center with restaurants, grocery stores, shops, and a hotel. There, they could secure clothing, food, and a place to stay.

As the two approached the trailhead, Gemma was startled by a large figure in the shadows. It was a hairy bipedal life form. She quickly raised her rifle before seeing it was a metal figure which resembled a large skunk dressed in a Park Ranger's uniform. It pointed at her with one finger. Above its head was a public service message which read, "Skitter says don't litter."

Rimor explained, "It's just a metal sign. It's an example of their art. There's no such thing as a Geminum-mephitis hybrid. It's a product of imagination."

"It's disgusting," Gemma answered.

Lights from the shopping center glowed above the horizon ahead.

The couple exited the woods into a clearing. What they came upon was appalling. Less than ten yards away was a parked automobile. It was purple and chrome. The thing had big wheels and a gaudy paint job. Two males stood beside it. They were large men dressed in what appeared to be casual suits. Two angry females were with them. One was large and wore a sour scowl. The other imitated her.

One of the men dragged a third woman across the asphalt pavement by her hair. Her knees were bloodied by abrasions. The young woman's nose bled. "No mas," she cried. "No mas!" A small child cried, "Leave my momma alone." The child beat on the leg of the man whom the others called "Junior." Annoyed by the kid, Junior was about to slap him down when he noticed Gemma and Rimor. In his drug-infested mind, they were two new toys for vicious amusement.

"Well, well, what have we got here?" he mused. Junior raised his head and looked down his slimy, cocaine-laced nose at them.

His two female accomplices taunted their frightened victim, who knelt hopelessly on the pavement. "What you say now, Chiquita?' growled the one called "BeJe."

"Guess what, b-itch?" she asked Chiquita. "I'm gonna scratch yo' face off with my keys. Ain't nobody gonna recognize you. They gonna call you a freak."

The other woman, named "Veejay," cackled loudly. "She gonna freak you, b-itch!"

Junior's discovery of Gemma and Rimor drew the women's attention. BeJe smacked the sobbing woman. "Chiquita" slumped to the ground. Veejay kicked her hard in the side. Then, they cast their attention to the Geminum couple.

"It ain't Halloween, homies," Junior sneered. "What we got here? Is you a space man? Well, it ain't Halloween, homey. They ain't no trick or treat here. Hu-uh, all you gonna get is tricks, my man. You know?" Then, he eyed Gemma. "So, what is you? Is you his space mechanic? Is you his little wrench monkey? Naw, you gonna be my wrench monkey tonight, girl. Now, why don't you climb outta them coveralls? You and me gonna party."

Gemma's anger grew. She wondered, "Were these the people who Rimor said were like us?" If so, she was badly disappointed. Prison on Gemina would be preferable to living in their horrid world.

BeJe added, "Look, they brung water soakers! Y'all gonna squirt your water guns at us? Ooh, we so scared!"

Veejay sneered at Gemma with her crooked, gap-toothed smile. "B-itch, I'm gonna f*ck you up."

They didn't understand Gemma was having a very bad day. If they didn't back off, it was going to be their bad day, too. Gemma stepped toward the women and raised her pulse rifle. "I don't know who the hades you people are, but this has been the worst day of my life. You really, really don't want to be part of it. Now, step back. Get away from that woman and child."

"Well skank," said BeJe, "yo' day gonna get a whole lot worser."

"Hey, Junior," Veejay yelled. "Who the hell is 'hades?' Is she disrespecting us? I think she's disrespecting us!"

Junior laughed, "Veejay," he replied, "That's exactly what they is doing. They come here and wanna order us around. Finish what you doing to Chiquita. Then, we'll play with the wrench monkey."

BeJe drew a five-inch blade from her blouse. "Watch this," she taunted Gemma. She grabbed the sobbing young woman's hair. Violently, BeJe dragged the crying woman across the raw pavement. "The woman pleaded, "No mas, por favor, no mas!" Chiquita's misery delighted her cackling tormenters.

With a cruel smirk, BeJe pointed her knife at Gemma. "You next, b-itch."

Gemma's anger was peaking. In a stern voice, she ordered BeJe, "Drop your weapon. Let go of the woman and stand back, now!"

The Ensign aimed her pulse rifle at BeJe's face. Rimor quickly cautioned her. "Gemma don't shoot. Your rifle is set at Level Two. You'll incinerate her. It will burn her victim, too."

Gemma didn't power down her rifle. Instead, she laid it on the ground. She had something special for the thug queen. BeJe taunted Gemma. "C'mon b-itch. Bring it. You know about yo' bad day? Well, b-itch, it's gonna get a whole lot worse. Bring it! What you got? I'll slash you, girl!"

Gemma approached BeJe. When they were an arm's length apart, she drew the Chief's wrench from her satchel. "This is what *I* got," she answered. In one powerful movement, she swung the heavy tool at BeJe's head. Instinctively, the woman raised her arm in defense. The wrench carried so much force that it broke BeJe's arm and slammed it into the side of her face. The crushing impact dropped her to the pavement. Blood flowed from BeJe's nose and mouth. Her arm was bent, broken. The fingers on her hand twitched. The only sound she managed was a low, gurgling moan.

Gemma stood above BeJe. "Welcome to my day," she scorned contemptuously. BeJe's victim, Chiquita, fearfully crawled away. Her boy followed her. The two hid and watched events unfold from behind the Hound's car.

BeJe's companion, Veejay, stepped away. She didn't want anything to do with Gemma. Veejay raised her hands and pleaded. "I ain't done you no harm. Don't hit me with that damn thing. I'm backin' away. Be cool, okay?"

Junior resented Gemma's interference with his playtime. He drew a pistol from his britches and held it sideways, aimed at her head. "F*ck you, bitch," he snarled.

Without hesitation, Rimor quickly drew and fired his pulse pistol at Junior. It was set to stun. The bully collapsed to the ground, where he lay semi-conscious, groaning.

The other male stepped forward. He was bigger and angrier than Junior. "You a big man with that space gun," he growled. "Space-boy, put the damn gun down. Face me like a man, or a *space man*. I don't know what the f*ck you are, and I don't give a damn. When I'm done with you, yo' momma won't know who you is or what you is. Then, I'm gonna take yo' b-itch and do her till she can't scream no more. She gonna get some Hound, you know? She gonna get some Hound while you lay dead by her head. Yeah, 'dead by her head!"

Rimor put down his pistol. It had been a very bad day for him, too. Soon, it would be Hound's bad day, probably the worst day of his miserable life. Hound pulled a six-inch blade from the back of his jeans. He approached Rimor, who unsheathed his battle blade. It was nine inches of razor-sharp death. Unlike Hound, the Lieutenant was well-trained in the use of *his* weapon. He nodded to let Hound know it was time to stop jawing. Rimor's blade would do his talking. It would welcome Hound to a very, very bad day.

Hound noticed how well Rimor's blade fit his hand. The blade flashed like silver fire with its every movement. Rimor motioned for Hound to fight. He would slash him apart.

Hound backed away. There would be no fight. "Nahha, man. Where'd you get that damn thing? What are you? Is you some sort of killer? Is that it? You a killer?" He knelt and slowly laid his knife on the ground. "I ain't givin' you no trouble, space man. Not now."

Rimor ordered the group to empty their pockets and handbags. "Everything on the ground, now," he commanded. They dropped keys, baggies, pills, pepper spray, cash, and credit cards. Junior's driver's license fell to the pavement.

"Where are the car keys?" demanded Rimor.

"F*ck you, space man," snarled Hound. "You ain't takin' my wheels."

"Throw the keys here, *now*." Rimor ordered. "Do it, or I'll cut them out of you." Gemma readied her pulse rifle. It was still set at level two. Her finger slipped into the trigger guard. Hound was within a microsecond of being incinerated. Every cell in his body would explode into micro-flames. He would be consumed by a fiery, massive chain reaction. Nothing but cinders and smoking ash would remain.

Reluctantly, the large man pulled a key fob from his pocket and tossed it to Rimor. "This ain't over," he snarled. "I'm gonna get you space man. Maybe not tonight, but I'm gonna get you and yo' b-itch. Nobody messes with Hound and lives." He nodded his large ugly head. "You a dead man."

BeJe wanted to trash talk Rimor and Gemma. Instead, she rolled on the ground helplessly. What came from her broken jaw, swollen lips, and cracked teeth sounded more like gargling.

Gemma gathered their cash and cards. Rimor walked towards the vehicle. He signaled for "Chiquita" and her boy to get in. They didn't know if they had been rescued, or if they had become his victims. The little boy cried, "Please help my daddy." The woman didn't enter the car. She pounded on the trunk lid. Rimor opened it to find a badly injured man. He was unconscious, sprawled in a pool of blood.

"He needs immediate medical attention," Gemma said.

The boy tried to crawl into the trunk with his father.

Gemma turned to Hound. "You did this?" she asked. She was furious.

"Yeah, and I'm gonna do it to you and Spaceman. But you gonna get some Hound first, b-itch. I'll find you space monkey. You'll beg me to kill you when I do. Ain't that right, Chiquita?"

"You really don't know much about women, do you? We have cycles. We can change from very nice to bad very quickly. Right now, my cycle is changing to very, very bad. Gemma raised her pulse rifle towards the man. With the slight squeeze of her finger, there was a bright, burning flash. Hound was no more. He was a twisted lump of cinders, some still glowing, sizzling. They made a slight crackling sound. Gemma looked down at his ashes. "Welcome to my day, Hound."

Rimor told her, "We can't move Chiquita's husband. He's hurt too badly."

"We can't leave him in the car trunk!"

"We'll drive him where he can get help." Rimor looked at Chiquita. "Come with us." She took her son by the hand. They climbed into the back seat.

"Gemma, give them the green paper bills. It's valuable, here. Chiquita can use it. Give them Junior's driver's license." Rimor pointed at it. "Law officers will use it to identify him."

Rimor drove across the parking area. Two hundred yards behind was Junior, BeJe, Veejay, and the smoldering ash that once was Hound. He parked Hound's purple pimpmobile in a well-lit area. Nearby was an idling patrol car. There was an officer behind its wheel, asleep. Rimor and Gemma exited. "Chiquita" sat behind the steering wheel. The trunk was open, so the police would see the boy's injured daddy. Its lid still bobbed up and down from their brief trip. "Blow the horn," Rimor called. He gestured so Chiquita would understand what had to be done.

Gemma and Rimor quickly left the parking lot. They entered the nearby hotel. Behind them was a loud, honking car and commotion. As it gathered attention, Rimor quietly rented a room. The hotel clerk was puzzled by the couple's attire. The Lieutenant shrugged his shoulders and said, "cosplay." The man winked and handed him a key.

In their fifth-floor room, Gemma looked from their window at several patrol cars. They were parked by Hound's wheels. Emergency medical personnel tended to the woman and her husband. They wrapped a blanket around the boy. "Chiquita," as the thugs called her, pointed where she had been attacked. One car sped there with flashing lights. It would find stunned Junior and battered BeJe. Veejay had already fled.

Gemma was exhausted. It had been a long, grueling, miserable day. She had worn the same outfit without sleep for nearly forty hours. "I'm dirty, Rimor. I need to get out of these filthy clothes. Do they have showers here?" Rimor nodded and pointed to the bathroom.

"Turn around," she asked, "I'm going to undress." As Rimor turned, she unzipped her coveralls, which dropped to the carpet. Gemma pulled off her sweat-drenched top. Her perky breasts bounced from their release. She wiggled free from her panties. They slid down her legs to the carpet. The room's air conditioner hummed softly and blew a cool breeze across her hot, sweaty flesh. It was refreshing.

Rimor intended to respect Gemma's modesty. However, an inadvertent glance at the dresser mirror caught her reflection disrobing. He was ashamed of violating her privacy. Before he could avert his eyes, the young Lieutenant noticed something disturbing. Gemma was bleeding. After their confrontation with Hound and company, he assumed she was wounded. He immediately turned to help.

There, Gemma stood before him undressed and surprised. "They hurt you," he said. "You're injured. Let me take care of that." Gemma looked down. She was bleeding. However, hers was unwounded blood.

Gemma was nonplused. In her culture, unwounded blood was sacred. The powers of creation and renewal journeyed therein. Once Rimor understood, it was a journey they would take together.

Somewhat amused, she understood Rimor's overreaction and allowed his approach. Unclothed and unashamed, she stood proudly before the young explorer. Gemma pointed across the room. Atop the bathroom counter were feminine napkins. "They have bandages there," she pointed. "Let me bathe before you tend me." Embarrassed, he quickly realized she was menstruating.

Gemma smiled. There was no doubt about Rimor's devotion. "I love you," she whispered in his ear. "Thank you for loving me." Her wounded heart bled no more.

[End of Chapter]

CHAPTER FOUR

Astraea's Promise

It was dawn on Gemina. The day began with a warming sunrise. The forecast was favorable. Few noticed distant thunder over the horizon. No one expected the oncoming storm. News of the attack against Outpost Semotus would soon break.

Loved ones gathered for a wedding. There were warm greetings, laughter, and joy. Spirits were high. Revelers were oblivious to the approaching conflagration.

Most attended because of the bride. Her name was Alexia Arabella Astraea. Astraea was an exceptional officer in Gemina's Galactic Navy. She graduated with honors at the top of her Academy class. Lieutenant Astraea served as Science Officer aboard the mighty *GSS Vindictus.* That was the planet's most powerful warship. No enemy could attack Gemina without first vanquishing the mighty dreadnaught. Then again, Gemina had no enemies, or did it?

Her groom was a local councilman named Apius Caecus Fidelis. It was customary on Gemina to call others by their last name, so most called the bride "Astraea" and knew her betrothed as "Fidelis."

Fidelis and his best man, Juvo, were late to the wedding. Juvo drove hurriedly to the Templum. He knew they were late and glanced repeatedly at his watch. As they sped to the Templum, Fidelis watched the sidewalks. His constituents waved happily when they saw him. He smiled and waved back, but not as a politician. The councilman truly cared about his constituents. To him, they were an extended family of sorts.

Astraea awaited in the Templum with her bridesmaids and family. The young couple were totally opposite. She was an extraordinary, beautiful young woman. The awaiting bride-to-be was tall, elegant, and graceful. She had a contagious sense of humor and empathy.

Unlike his bride, there was nothing glamorous about Fidelis. He always wore plain, civilian suits, even to his wedding. His aspiration was to be a humble public servant. Fidelis and Astraea may have lived on the same planet, but their lives traveled paths far apart. There was no place for Fidelis in the stars traveled by Astraea. Few understood their love, genuine as it was.

Fidelis and Juvo were minutes from the wedding. Suddenly, a boy darted across the road in front of them. The child carried a lady's handbag. Juvo slammed his sedan's brakes to avoid hitting the fleeing delinquent. He narrowly averted a tragic accident.

An angry woman pursued the boy. "Stop him," she frantically yelled. The young thief was about ten years of age. His clothes didn't quite fit. They were soiled and ragged. Juvo stopped his vehicle. Fidelis quickly exited.

Juvo yelled to his friend, "Hey, Fidey, Get back in the car. We got a wedding waiting for you. We got to go, now!"

Fidelis answered with his usual quick smile, "Give me a minute. I'll be back."

"Please, Fidey," worried Juvo. "You're going to be late to your own wedding! Hurry! If we don't get there soon, she might marry an Admiral!"

"Stay here," Fidelis told his friend. "I know this kid." He pursued the fleeing youth to a dead-end alley. The councilman approached the child. The boy stepped back. He had nowhere to run. With a panicked scowl, the boy drew a kitchen blade from his pocket and flashed it at the young councilman. Fidelis called to him, "Listen to me, Puer. Put the knife down."

The child shook his head. "I can't."

Fidelis patiently gestured to the boy, "Hand it to me, Puer."

Fidelis stepped forward cautiously. Puer slashed the councilman's jacket sleeve. Puer raised his blade shoulder high and threatened to strike again. The councilman knelt to one knee. He peered compassionately into the child's eyes and extended an open hand.

"Give me the knife, Puer."

There was an uneasy pause. The boy looked from side to side. He searched for a quick escape.

There was none.

Puer lowered his head. He dropped his arms to his sides and began crying. The two shared the moment, much as would an older and younger brother. Fidelis took the blade from Puer and put it in his jacket pocket. In a calm voice, the young councilman asked, "What's going on, Puer? What's wrong?"

The child sobbed, "Daddy left. We don't know where he's at. Momma can't pay the bills. My sister ain't complaining, but she's real hungry, Mr. Fidelis. I know she's sick and hurting! We got no money, no food, nothing. I'll pay the lady back. I promise! I don't want my sister to be hungry no more. I don't want momma cryin' no more. Please let me go."

"Give me the pocketbook," Fidelis replied.

"You don't understand, Mr. Fidelis. Our house is dark and cold. The refrigerator is empty. We ain't got no water, nothing! Momma and my sister are scared! They got nobody!"

"Give it to me, son."

Hesitantly, Puer handed Fidelis the pocketbook. The child's face was tight from distress. His eyes were red, watery. Fresh tears dried on his cheeks.

Fidelis pulled a wallet from his torn wedding jacket. He drew every bill from it and handed the money to Puer. "Take this to your Momma. It will help. You, your momma, and sister can have supper tonight. All I ask is that you thank the lords for watching over you."

"I didn't want to admit it, Mr. Fidelis, but I'm hungry, too. Thank you," he sobbed.

"Puer, does your mom have a telecom (advanced telephonic device)?"

"No, sir. She ain't got nothin' or nobody."

"That's not really true. She has you." The councilman removed his own telecom. He handed it to Puer. 'Give your momma this." Reluctantly, Puer accepted. Then, Fidelis gave the boy his card. "Have her call this number tomorrow. Tell them I asked her to call. They'll help, okay?"

"What about your telecom, Mr. Fidelis?"

"Tell your Momma to keep it for now. Go! Your family is waiting for you."

"I'm sorry, Mr. Fidelis. I didn't know what to do."

"Listen to me, Puer. There are times when none of us know what to do. That's when faith guides us. Everything will be alright. Just don't do anything like that again. Okay? If you ever need help, just come to me." The boy hugged Fidelis. Then he ran home.

Fidelis handed the pocketbook to its panting, short-winded owner. She thanked him. "Thank you, Mr. Fidelis," she said. "I got bills to pay, too. I'm sorry for that child. I would have helped him if I could. But I couldn't do what you done. Bless you, sir." The two bade farewell.

The councilman returned to his ride. Juvo was concerned. "Get in, Fidey! We're late for your wedding!" Then, Juvo paused. "Hey Fidey, the kid slashed your wedding jacket! You're not only late for your wedding, but you're gonna get married looking like you've been in a bar fight!"

Fidelis looked at his sleeve. He shook his head and grinned. "I don't think Astraea would have me any other way."

Juvo laughed. "Hey, the kid gave you a knife. Maybe you can use it at the wedding?" Fidelis answered with a stern look. Juvo added, "I'm just sayin! I mean Astraea has one. She gets a battle blade."

The groom-to-be laughed. "That's funny Juvo, but Astraea won't use her knife to fight me… At least, I hope not."

"I guess not," Juvo admitted. "You know?" he asked. "I can't figure it out. I mean nothing personal, but you and Astraea are the oddest pair. She's this goddess, and you're like a gardener! The gods must be pranking her. I hope she has a good sense of humor!'

"Shut the car door," Fidey. We got to hurry to the Templum before your bride realizes she's making a huge mistake!"

Fidelis assented with a slight nod. "Juvo, I thank the lords that love is blind."

"She does see you, councilman. We all do."

At the Templum, Astraea patiently awaited her betrothed. Her sisters and maidens gathered by her side. They fretted about Fidelis being late.

"Where could he be?" the bride's youngest sister asked impatiently. Astraea's parents glanced nervously towards the Templums' doorway.

Astraea answered with a smile. "He's probably on one of his adventures."

"Fidelis doesn't have adventures," her oldest sister protested. "He's a civilian! An adventure for Fidelis is leaving a budget meeting early." She shrugged. "He battles with words and gestures. Frankly, sis, I don't understand what you see in him."

Astraea kissed her sister's cheek. "Would you have me marry a powerful soldier? Is that it?"

"No, we want you to marry for love. But how can you love Fidelis? He isn't anything like you. He's no hero!"

"Sisters," she explained. "Fidelis owns power differently. He battles with his heart, not his fist. Look upon me. See my happiness. He has won me, not conquered."

Her little sister giggled and answered, "I know why Sis is sweet on him. Fidelis is eye candy!"

For a moment, Astraea fondly recalled her first meeting with Fidelis. They met at a restaurant, very much by accident. They shared lunch, talked, laughed, and enjoyed the other's company. Lunch turned into a day, then another. They were together for weeks, then months. Soon, it would be forever. She and Fidelis had an undeniable bond. From their first day, they were inseparable.

Astraea's remembrance was interrupted by her youngest sister's advice. "Astraea, when he gets here," she urged, "you should give him one of those knockout punches they teach at the Academy!"

The other sister added, "Yeah, Sis! You need to show him who's boss!"

"When he gets here," Astraea told them, "I will give Fidelis my love, and the kiss which shall join us as husband and wife."

It was then that Fidelis and Juvo arrived.

"He's here," the bridesmaids called excitedly. The groomsmen hurriedly rushed outside. They guided the groom towards the altar. As Fidelis entered the Templum, he saw Astraea. He would never forget how calmly she stood in the distant archway with her young maidens by her sides. Astraea was so stunning that Fidelis lost his breath.

She proudly wore her white dress uniform and a silver, plumed Cuirassier style helmet. White lace flowed from its gold embossed crista. She stood in white, high-heeled boots. The young Lieutenant appeared taller than her groom. Astraea held her head high, shoulders swept back. Her every movement shed grace.

Pure white-laced armorettes adorned her arms. Her hands were accented by elegant, long-fingered gloves. No goddess would have been more impressive.

The gathering chanted. They wanted Astraea to journey down the aisle and take her betrothed's hand. Fidelis stood in a daze. Astraea's dazzling eyes sparkled crystal blue in the light, then turned hazel green in shadow. Her emotions were often reflected by the hue of her gaze. Whenever she looked upon Fidelis, her hue was green.

The groom glanced into the crowd. He saw his family and friends, smiling and clapping. On the other side of the aisle were soldiers, stellarii, and other servicemen. All were smartly attired in dress uniforms. They were an impressive sight. Astraea's party stomped their feet in unison. That was a common military gesture of support to a comrade – in this case Astraea.

Then, his groomsmen walked Fidelis to the altar. The groom's knees wobbled slightly. His torn jacket sleeve was noticeable. Many figured it evidenced another of the young Councilman's adventures.

The Priest quickly stepped to the groom's side and steadied him.

As the gathering chanted and cheered, Astraea ceremoniously removed her helmet with both hands. She placed it upon a silken pillow held by one of her maidens. Her long, flowing brown hair dropped below her shoulders, and swirled with the turn of her head. Two blond streaks accented each side. The remainder was rich, shimmering. Such coloring was common to those from her Province. Yet, it was extraordinarily elegant.

The young maiden who carried Astraea's helmet stood quietly. Astraea slowly drew an ornate, silver and gold battle blade from her thigh sheath. She raised it above her head with both hands. Everyone watched as she lowered the weapon onto a white silk pillow carried by her other maiden. Having disarmed herself, the bride stepped forwards to accept the fates of marriage. She looked ahead and met Fidelis' gaze with a bright smile that melted his soul. They were so much in love that nothing, nor anyone else, had their attention.

The bride took a long step forward, followed by her bridesmaids. The gathering began to clap in unison. Women cried and sobbed. Astraea took a measured second step, then a third. Ceremoniously, she approached her beloved. Standing before their families and friends, the couple would soon be joined as husband and wife.

But it was not to be.

Without warning, the wedding was interrupted by the howling sirens of emergency vehicles outside. Bright, flashing red and blue strobe lights bounced against the building's stain glass windows. The inner Templum changed hues with each strobe flash. A group of Marines burst inside. They wore light, drab olive-green battle armor. Each carried an automatic pulse rifle. One called out, "Lieutenant Astraea, you are needed immediately. Please come with us, *now*."

Astraea realized it was a call to duty. A crisis was at hand. She raced to Fidelis, embraced him, and gave him the kiss that would have wed them. Astraea said softly, "I love you more than anything else, but I must leave. I promise to return. Please, wait for me."

Before he could reply, his bride turned and left with her Marine escort. Silence overtook the Templum. Servicemen in the gallery rose and promptly exited the building. Others sat quietly, stunned. Everyone knew something very bad was happening. There were murmurs from some, "It's war!"

[End of Chapter]

CHAPTER FIVE

Call to Duty

Astraea answered the call to war. She left her family and friends at the Templum. Most hurried home. Others reported for duty. Fidelis didn't know how to react. He remained at the altar. The Priest stood beside him. "Have faith, young man," offered the cleric. "Your bride will return."

The Councilman stuttered, "I feel helpless, hopeless, Father. Astraea's gone to war. There's nothing I can do."

The clergyman knelt at the altar. "My son, would you join me in prayer?"

Fidelis accepted the offer.

"Dear lord, great thou art, we beseech you to guide and protect young Lieutenant Astraea. Cover her with your holy shield and protect her from harm's way. Bring her home safely to her betrothed, to young Fidelis who prays at my side, and to her family, and to we others who love her.'

Fidelis fought back tears. He added, "My Gods, please don't let anything happen to her."

As he unclasped his hands, the Priest looked to the heavens and concluded, "Amen."

Fidelis confided, "I know she's gone to war, Father. I know she's far away, but I feel her love." He added, "Our hearts beat as one."

"Her spirit is within you, my son. Yours is with her. It will make her stronger."

"Father, I would gladly take her place if I could. If I were a soldier, I would fight beside her. I would do anything to protect her!"

The Priest nodded. "There are others, such as yourself, who feel helpless. Please understand my son. In troubled times, the lord allows us faith. It is not given. We must seek it and help others find it. That is what your bride wishes upon you. That is your duty to her."

Fidelis wasn't sure the Priest understood his plight. "Father, Astraea is a warrior. Death is no stranger to her. It pretends to be her companion. I fear for her life."

The Cleric rose. He put his hand on Fidelis's shoulder. "Indeed, you must. However, Astraea calls upon you to go to others. Comfort them. Help them find strength for what follows. That is where your battle must be fought."

Fidelis stood. He gathered his courage. The young councilman left the Templum. Juvo waited outside. They embraced. Then, Juvo drove his best friend home. At the doorstep was Puer's mother. In her frail hands was a bowl of fresh, steaming stew. She made it with groceries purchased from money the councilman gave her son. "We prayed for you, Mr. Fidelis," the matron said. "The lord will protect you and yours, as you protected my children and me." She handed him her offering, then hugged the young councilman and kissed him on the cheek. Her shy, young daughter tightly grasped the woman's arm. She looked upon Fidelis with a bright smile and waved goodbye. "Thank you, Mr. Fidelis."

Juvo looked at the stew. Then, he glanced at his friend. "Hey, you know I'm hungry, too! How come nobody brought me anything to eat?" Fidelis took him inside and handed Juvo a spoon. The men shared grace and a delicious meal. They chatted for fourteen minutes. Once he was satisfied that Fidelis had regained his composure, Juvo returned home to his wife and children. They anxiously welcomed him.

Fidelis's constituents gathered in the streets. Many were crying, worrying. He shared their suffering and despair. The young councilman asked each to stand and speak about their loved ones. One after another, they did so, giving voice to their love and faith. At first, they sobbed and cried, but after telling their stories, many in the gathering laughed, cheered, and rejoiced.

Fidelis asked them to pray for one another. They did so. Many held hands. No longer strangers or acquaintances, they were a community. They dispersed with newfound strength. Faith gave them reason for hope.

That was what the *Vindictus* crew needed. It was why they fought.

[End of Chapter]

CHAPTER SIX

Faster Than They Look

Three Marines rushed Lieutenant Astraea to a nearby DUK transport. It was near the Templum with its engines whirring. A well-muscled petty officer met the Lieutenant yards from the transport's ramp and saluted her. "Petty Officer Iratus reporting, ma'am."

"Are you assigned to the *Vindictus*, petty officer?" Asked Astraea.

"No ma'am. I'm just looking for a ride."

"The *Vindictus* is not a passenger ship, Mr. Iratus."

"No, ma'am."

"So, what's happening, Petty Officer?"

"We're at war."

"With whom?"

"I don't know ma'am. All I know is that Outpost Semotus has been attacked."

"Are you sure?" Astraea asked. "Semotus is in deep space. It's near the expanse. There's nothing out there."

"Yes, ma'am. The Outpost is under attack as we speak. Fleet Command dispatched all available warships. All personnel have been called to duty."

Astraea knew Gemina had few warships. The planet saw no reason for a large galactic battle fleet. It had no known enemies. Gemina had never before been attacked.

Gemina's Navy had fifteen ships. They were built to deal with smugglers and commerce raiders. Included were six destroyers, eight frigates, two powerful battle cruisers, and the Navy's only dreadnaught, the mighty *GSS Vindictus*.

Astraea asked Petty Officer Iratus, "How many enemy ships are attacking Semotus?"

"There are a lot of them, Lieutenant. That's all we know."

"What classes of ships are we facing?"

"That's uncertain, ma'am. Reports from Semotus said they have interceptors, destroyers, and battle cruisers. They may have dreadnaughts. No one knows for sure. They're murdering the Outpost, Lieutenant. We've got to do something."

Astraea nodded her head in angered agreement. She replied, "We certainly do, Mr. Iratus. There are more than twenty thousand people on Semotus. There's no justification for any attack against them. The Outpost trespasses on no one. It's in an unpopulated sector of space without military value. The only thing the enemy can expect from their attack is a counterattack."

"Yes ma'am," answered Iratus. "I have buddies and crewmates there. They need help."

An older man inside the DUK began screaming. He burst through the hatchway and sprinted down the ramp past the Lieutenant. Dressed in a flag officer's uniform, he repeatedly yelled, "Frack this, frack this!" The runner sprinted frantically towards the parking area. "No fracking way!"

Iratus hollered at him, "Hey, Admiral, the War's that way." He pointed above his head. The Admiral didn't care. He kept running. Two marines made chase, but the flag officer was faster. He boarded a nearby vehicle and fled towards town.

"Was that Admiral Pavidas?" asked Astraea.

"It appears so, ma'am. He's faster than Captain Pullum! Them senior officers can fool you, Lieutenant. They're a lot faster than they look!"

"Tell me, Iratus, is the *Vindictus* ready for war?"

"Ma'am, I don't think anyone is ready, but most of the crew is aboard and ready for duty. The ship needs more gunnery officers."

Another serviceman ran shrieking past them. "No frackin' way, No, no, no frackin' way!" he yelled hysterically. The same two marines gave chase. After a lengthy pursuit, they caught him and dragged the flailing man back to the transport.

Iratus pointed to the panicked man. "That's your ship's gunnery officer, ma'am. I ain't as fast as him, but I know my guns. I ask for permission to join the *Vindictus*, Lieutenant."

"Granted, Petty Officer Iratus. The *Vindictus* has many guns. I'm sure we can find one for you."

The two boarded the waiting DUK. Minutes later, they were aboard a launch docked with the *Vindictus*.

[End of Chapter]

CHAPTER SEVEN

Captain On Board

The *Vindictus* was the most powerful Geminum warship ever built. She had eight graviton guns, twelve automatic pulse cannons, and a variety of other weapons. Perhaps, more formidable was her crew. They were the best in the Fleet. Lieutenant Astraea was an example of the ship's elite crewmen. However, no one could expect the duties thrust upon them that day.

Once the Lieutenant's launch had docked with the *Vindictus*, she and Petty Officer Iratus boarded the dreadnaught. As the launch doors opened, they stepped aboard the Bridge. The Lieutenant, of course, was first. To her surprise, a bosun's whistle sounded. The crew stood at attention. They turned towards Astraea and saluted. "Captain on Board," announced Ensign Carpo. Astraea turned to salute Captain Timens, but he wasn't there. Iratus was the only person behind her. She then realized the crew was saluting her!

Astraea was the Captain! Nonplused, she returned their salute and addressed the bridge crew.

"Lieutenant Telum," queried Captain Astraea, "Is Captain Timens not aboard?"

"No, ma'am. He was planetside with the Executive Officer. They were in a Launch accident. Mr. Fugiens died from his injuries. Our Captain was badly wounded. He's hospitalized. We have no senior officers aboard."

"Mr. Telum, give me the ship's status report."

"The *Vindictus* is battle ready, ma'am," answered the weapons officer. "Propulsion systems are engaged. Navigation is online. Pulse cannons are ready. However, ma'am, one graviton cannon is unmanned. We could use another gunnery officer."

"Petty Officer Iratus, are you familiar with graviton cannons?"

"Ma'am, I'm familiar with every gun in the fleet."

"That cannon is yours, Mr. Iratus. Please man it now."

"Aye, ma'am." Petty Officer Iratus hurried to his new post.

"Ensign Carpo, have we received reports from Fleet Command?"

"Aye, Captain. We're receiving them now."

During the next five minutes, the Ensign relayed accounts of the Battle of Sector 16. Some referred to it as the "Battle of Semotus."

Carpo reported, "Admiral Forctis is leading a squadron of Geminum warships into Sector 16. Their mission is to relieve Outpost Semotus from attack by unidentified warships. The Admiral commands eight vessels. That includes both our battle cruisers, the *Essex* and *Renown*, and two G Class Destroyers, the *Lioness* and *Tigress*."

"Are you certain of that Miss Carpo? Admiral Forctis retired last year after serving seven years as Commandant at Virtus Naval Academy."

"Aye, Captain. They said the Admiral volunteered to lead the squadron. No other command officers were available. He was the only one to board a transport."

Carpo pressed her earpiece closer to hear transmissions. "Captain Astraea, the Admiral's squadron dropped out of FTL. The *Reserve* reports that Outpost Semotus has been demolished. Everything is crushed and burning. There appear to be no survivors."

Grave concern swept across the bridge.

"The enemy continue to rake Semotus with bombs and cannon fire, although there are no standing buildings or moving vehicles. There's nothing left of the Outpost."

"Captain, our squadron has engaged the attackers. The *Tigress* targeted one of their destroyers."

"Excellent," replied Astraea. The *Tigress* is a powerful warship commanded by Commander Evenan Titos Scitus. He's an excellent tactician."

"No ma'am, Commander Scitus was unable to board the *Tigress*. His launch was damaged in the same accident which killed Mr. Fugiens. The *Tigress* left without him."

"So, the *Tigress* is skippered by a junior officer?"

"Aye, Captain. I'm getting a new report. The *Tigress* made direct hits on the enemy vessel's superstructure with six gigawatts of pulse cannon fire."

The bridge crew began to cheer, until Ensign Carpo continued, "without effect. The *Tigress* has moved on to another target. It appears to be a frigate. Several enemy ships have engaged her, Captain. The *Tigress* has been struck by multiple salvos of cannon fire. Her shielding is failing. She's being hit everywhere."

Carpo listened intently. She reported, "Captain, the *Tigress* suffered multiple hull ruptures. She imploded and is lost. There are no survivors."

Surprised, Astraea queried, "Ensign Carpo, could that be mistaken? The *Tigress* is a Class G Destroyer with a vitaelium hull and quantum citadel. Are you certain of your report?"

"Aye, ma'am, the *Tigress* was destroyed."

"Commander Telum, can you confirm that?"

"Aye, Captain. The *Tigress* suffered multiple cannon blasts. Her shielding provided little protection. She suffered catastrophic hull damage."

Astraea directed the Ensign to continue her reports.

"The *Renown* has engaged a hostile cruiser. Her main turrets have targeted its bow and stern. She unleashed a full broadside on the enemy, without effect."

One crewmember quietly commented, "None of our weapons have any effect!"

Captain Astraea glanced at him with disapproval. The battle was not going well. The *Reserve* and *Tigress* had been lost. No enemy ship was damaged. Not even the powerful *Renown* injured them.

Captain Astraea leaned forward. Everyone was quiet. She directed, "Miss Carpo, continue your report."

"Captain," the communications officer reported, "the *Renown* is receiving fire. She has suffered hits to her transom, stern, superstructure, forecastle, and bow.'

The Ensign paused after receiving further information. "Ma'am, the *Renown* has been slashed apart and is lost. There are no survivors."

There was shocked silence among the *Vindictus* bridge crew. Mr. Telum murmured, "That was the Admiral's flag ship! They're wiping out our entire squadron."

Astraea was shaken. The Galactic Navy was a small community of fifteen ships. Admiral Forctis had taught Astraea and other midshipmen at the Academy. She lost friends and classmates aboard the *Tigress* and *Renown*. More would be lost in the battle.

Not even a battlecruiser could damage the enemy. That didn't deter the remaining Geminum vessels. They were determined to fight.

Miss Carpo reported. "Captain, the *Essex* is intercepting an enemy cruiser. She has landed several hits on the ship, including a concentrated broadside to their superstructure. The enemy vessel is undamaged. It has been joined by two other warships. They're pounding the *Essex*."

Mr. Telum commented, "The Essex is commanded by Captain Pullum. They may think they're getting the best of him, but the Captain will have a trick or two for them."

"No, sir," Miss Carpo replied. "Captain Pullum isn't aboard. He didn't report for duty. The *Essex* is suffering extensive damage. Her CMS has been knocked out. She's fighting blind. Her aft hull has buckled. The ship has lost propulsion. She's leaking plasma." After another pause, Ensign Carpo added, "Captain, the *Essex* has lost power on six decks. Several enemy destroyers have joined the engagement. She has taken several more hits to her forecastle." After a pause, Miss Carpo reported, "The *Essex* has disintegrated. There are no survivors. She was lost with all hands."

Tears slipped down the navigator's cheeks. She sobbed, "My sister served aboard the *Essex*. They killed her."

Everyone had already lost someone. More would perish.

Carpo continued. "Captain Astraea, the *Invincible* is leading an attack on a large enemy vessel. It's an interceptor carrier. She targeted its fore and aft elevators.

The *Invincible* is pounding the enemy ship's command island and flightdeck with her main batteries, without effect. Two enemy destroyers are engaging *Invincible*."

Miss Carpo continued. "The *Invincible* is maneuvering to unleash another salvo against the carrier. She's been hit." Carpo listened to Fleet Command's report before continuing. "Captain, the *Invincible* has suffered multiple hull breaches and is bleeding plasma into space. Her propulsion systems are compromised." The Ensign hesitated before continuing, "There are reports of a violent fusion eruption. The *Invincible* is lost. There are no survivors, ma'am."

"Get those ships out of there," urged Astraea, but she had no command over them. "Miss Carpo," she asked, "who is the squadron's senior officer?"

"I can't say, ma'am. Several ships are skippered by junior officers."

"Like us," answered Astraea.

"Captain, the *Inflexible* and *Indomitable* have engaged three enemy warships. They landed multiple cannon strikes against them, but their weapons were ineffective. The frigates are taking fire, ma'am. They've been damaged.'

"Captain, we're receiving a transmission from the *Lioness*. Commander Clades has taken command of the squadron. He ordered the *Inflexible* and *Indomitable* to immediately withdraw to Sector 14 at FTL. The *Inflexible* has suffered several more hits, as has the *Indomitable*." After another pause, Miss Carpo continued, "Both ships have disengaged. The enemy is pursuing them. The *Indomitable* has made FTL."

"What about the *Inflexible*?" asked Astraea.

"Her propulsion systems are compromised. She can't make FTL. Damage control crews are doing everything possible to get her engines back online. Enemy warships are closing on her. She is within their range."

The *Vindictus* bridge crew listened intently. Their faces were grim. The Geminum squadron had been reduced to three warships. Two of those were badly mauled.

Captain Astraea replied, "The *Lioness* is captained by Commander Jedrik Aegeus Clades. He's an experienced war veteran. It appears Mr. Clades is finding chinks in their armor."

"Ma'am, the *Lioness* blasted an enemy cruiser mid-ship and disrupted its power generators. The cruiser is dropping out of formation."

"Good work, Commander Clades." Astraea commended. "The *Lioness* has taken big bites out of two of their ships."

"Captain, the *Inflexible* has repaired her propulsion systems. She's launched into FTL. The *Inflexible* has joined the *Indomitable* in Sector 14."

Astraea commented. "That's where Admiral Forctis had gathered the squadron. He organized the task force there."

Miss Carpo continued. "Only the *Lioness* remains in Sector 16. Eight enemy warships are engaging her. She has taken a direct hit midship and suffered a hull breach. There are casualties. More ships are targeting her." Carpo paused before reporting, "The *Lioness* has escaped to FTL. She's in Sector 14."

There was a sigh of relief aboard the *Vindictus* bridge.

"Miss Carpo, are the enemy warships following the Commander's squadron into Sector 14?"

"It appears so, Ma'am. They've also gone FTL. The hostiles are pursuing the *Lioness*, *Inflexible*, and *Indomitable*. Commander Clades is leading his squadron into Sector 18."

Astraea commented, "There's nothing there. The Commander is leading the enemy away from us. He intends to give us enough time to establish a defensive perimeter."

Miss Carpo reported after receiving another transmission. "Lieutenant Astraea, it's Fleet Command. You've received a battlefield promotion to the rank of Captain. Congratulations, ma'am."

"Thank you, Ms. Carpo. Please hail the *Lioness'* Captain. Patch us to the *Indomitable* and the *Inflexible*."

"Yes, ma'am."

"This is Captain Astraea aboard the *Vindictus*. Commander Clades, I order you to immediately withdraw your ships from Sector 18. Rendezvous with us in Sector 12."

After a moment of hesitation, the Commander replied, "Captain, Sector 12 is a Red Zone. Space travel is prohibited there."

"I'm aware of that, Commander Clades. If we weren't at War. I wouldn't order it. Lead your squadron to Sector 12 without delay."

"Captain," replied Clades, "Please reconsider. The enemy are in pursuit. If we rendezvous in Sector 12, they will follow us to your position. I request you allow us to remain in Sector 18, where we can draw them away from you.

"You have my orders, Commander. Make flank speed to Sector 12. That includes your whole squadron. We have transmitted our coordinates to you."

Commander Clades was confused by the young Captain's command. Naval doctrine dictated him to divert a superior enemy away from his fleet, which would allow time to maneuver into favorable positions for combat.

Outpost Semotus was defenseless. Everyone there was killed. The Commander hesitated to obey Captain Astraea's order. She was young and inexperienced. Her command to leave Sector 18 for Sector 12 made no sense to him. However, hers was a military command. If Clades challenged her authority, disarray and confusion would spread within Gemina's remaining Navy. He would not allow that. Against his better judgment, Commander Clades followed Astraea's order not to engage the "Visitors."

"Aye, Captain," confirmed Commander Clades. "We're altering our present course. The *Lioness*, *Indomitable*, and *Inflexible* will join the *Vindictus* in Sector 12."

"Thank you, Commander," Astraea acknowledged. She then addressed Mr. Telum. "Sir, you are promoted to Lieutenant Commander, which is consistent with your duties. You will serve as my Executive Officer."

"Aye ma'am."

"Are any other ships joining us?"

"Aye, Captain. Three frigates are coming. They are *Repulse*, *Courageous*, and *Furious*. The hospital ship *Clementia* follows. She is accompanied by the *Defiant*."

"Direct them to rendezvous with us in Sector 12."

Ensign Carpo hesitated. "Captain Astraea, Safety Directives require me to inform you that Sector 12 is a 'No-Go' zone. Without exception, ships are not allowed there. The Sector is too hazardous for navigation. It is designated as a 'Red Sector."

"Ensign, let the record reflect that you informed me of that matter. Hear this, however, I won't repeat my orders. Have all ships proceed to the following coordinates." Astraea punched in the location. She further directed, "Ensign Carpo, you are hereby promoted to Lieutenant. Carry on."

"Aye, ma'am." Lieutenant Carpo then transmitted to the fleet, "All ships must rendezvous at the following coordinates in Sector 12."

"Mr. Telum, lay our course. Captain Astraea inputted another destination. Engage at flank speed."

Lieutenant Commander Telum didn't understand. "But ma'am," he protested. "Those coordinates will place us in orbit around planet Immanis. It's a heavy, gaseous giant. The *Vindictus* won't be able to maintain an orbit there. We'll be pulled into the planet's gravity well."

"I'm aware of that, Mr. Telum. Lay our course, now."

"Aye, ma'am."

"Lieutenant Carpo," directed Captain Astraea, "instruct all department heads to meet me in the Operations Room, immediately."

"Aye, ma'am." Carpo opened the ship's intercom and announced, "All department heads report to the Operations Room immediately. I repeat, all department heads report to the Operations Room. That is an order."

Lieutenant Commander Telum escorted Astraea there. He explained that every department head was a junior officer. "Ma'am, the *Vindictus* was ordered into battle without any senior officers aboard. All department heads are Lieutenants. They're unaccustomed to their new duties, but each is performing well.

"Mr. Telum, from what I saw planetside, some senior officers didn't want to be here. Perhaps, they believed we 'junior officers' were better suited for war than they?"

The two entered the Operations Room. Awaiting them were the ship's Lieutenants. Each stood at attention and saluted Captain Astraea.

"At ease," she told them. They took their places at the conference table.

Lieutenant Negatio rose. He was about to speak, but Astraea's cold, crystal blue eyes intimidated him. He stuttered, "Permission to be heard, ma'am?"

"Granted, Mr. Negatio."

"Captain, I'm in charge of damage control. Please allow me to offer you my counsel as its department head,"

"Please do, Lieutenant."

"With all due respect, ma'am, I don't think you should be our Captain."

"Mr. Negatio, you're aware I was assigned command of this ship. Would you have me shirk my duties?"

"No, ma'am. I suggest that Mr. Telum is more qualified to lead the *Vindictus* into battle. With due respect, you're an astrophysicist. All of us know you're brilliant, but we're not conducting scientific experiments. The *Vindictus* is going to fight an invasion force. Mr. Telum is the ship's Weapons Officer. He is familiar with our weaponry. Telum should lead us into combat."

"Please be seated, sir."

Lieutenant Figo rose. "Captain, may I have permission to speak?"

"Please do, Mr. Figo. What is your counsel, sir?"

"I'm in charge of engineering. It's an important job. I just want you to know that we've got a good engineering crew. We'll give the ship our best."

"Thank you, Mr. Figo."

He continued, "Ma'am, with due respect, I also have concerns. The *Vindictus* is Gemina's best chance of victory. We must fight an enemy of vastly superior numbers. Prime

Minister Morus calls them 'Visitors.' Whatever they may be, those things have already obliterated Outpost Semotus. They've damaged or destroyed more than half our fleet. Perhaps, Mr. Negatio is correct. Maybe, our weapons officer should command the ship in combat?"

"Mr. Figo, I've been promoted to the rank of Captain and given command of this ship. It is my duty to command the *Vindictus*. I tell you, sir, we will stop the 'Visitors' in Sector 12. We will kill them all."

"Captain," continued Figo, "I'm also concerned about Sector 12. It's a 'No-Go' zone. Those are extremely hazardous places. Must we add more dangers to a perilous mission?"

"I hear you, sir. I understand your concerns, but we will fight the enemy in Sector 12."

Mr. Negatio arose again. "Captain. Mr. Figo raises valid concerns. I have another. You intend to put the *Vindictus* in orbit of a gaseous giant. That's a mistake. The planet's massive gravity will stress our hull. It could break us apart. It will strain our propulsion systems and limit our maneuverability in battle. We could be drawn into a gravity hole."

"Mr. Negatio, what you say has merit," Astraea acknowledged. "I understand your concerns."

Dr. Medens arose. "Ma'am, I'm just a doctor, little more than a resident. As with my fellow officers, I have duties for which I'm unprepared."

"I understand Doctor, but no one is ever ready for war."

"Of course not, Captain. Please forgive my inexperience. I understand we're headed directly into combat. The *Vindictus* has an empty med bay now, but when we engage the enemy, it will be filled with injured and dying crew members. I speak to you as their physician. May I be heard?"

"I will hear you, sir."

"Captain Astraea, as you know, I'm not a military man. I may be saying this incorrectly." The Doctor paused. He felt uneasy about his inquiry. "Ma'am, are we going into Sector 12 to hide from the enemy?"

"Sir?"

"Many crewmen believe you're taking us into Sector 12 to hide from those things."

"Do you think we should hide, Doctor?"

"Ma'am, I'm just a physician, not an Admiral. It doesn't matter what I think. However, I speak for the crew on one thing. We don't want to hide from the enemy. We're the only hope for our families, loved ones, and our world. We want to fight."

"Thank you, sir. I assure you that we're not going to hide in Sector 12. We go there to destroy the enemy. Gentlemen, thank you for your counsel. You have your orders."

Negatio angrily returned to his feet, "Captain, we ask you to explain those orders."

Astraea answered, "A Captain's role is to give orders, not explain them. A crew's duty is to follow orders, not question them. We will soon engage the enemy. I depend upon you to follow my commands immediately without question or hesitation."

She continued, "Gentlemen, you carry responsibilities beyond your rank. Because of that, I promote each of you to Lieutenant Commander."

Astraea measured her officers' composure. They were uneasy, but that's expected of men facing combat. "You're dismissed," she told them. "May the gods be with you. May they be with us all."

The Captain returned to the bridge with Mr. Telum and Lieutenant Carpo. The other officers remained for further discussion. Negatio was visibly upset. He tightly gripped the edges of the conference table. The frustrated damage control Chief leaned forward and argued. "For the sakes of our gods, any god, we can't let that woman command this ship."

"She's the Captain," replied Mr. Figo.

"She's a Lieutenant, as we are. Astraea is an astrophysicist! What does she know about commanding a dreadnaught in battle?"

"Mr. Negatio," answered Figo. "Astraea was promoted to Captain. May I also remind you that none of us have led a dreadnaught into war. We have no more combat experience than her."

"Please Gentlemen," Negatio continued. "We're outnumbered by a vicious, swarming enemy. We need a Captain whose judgment we can trust. Nothing she proposes makes sense."

"May I remind you, sir, that Commander Clades has followed her orders. He's no junior officer. Mr. Clades is a respected veteran officer who commands a squadron of warships."

"Clades is a fool," Negatio replied. "He followed the Prime Minister's crazy order to leave Semotus undefended. That idiot got everyone in Sector 16 killed."

Dr. Medens replied, "I tell you, Mr. Negatio. Be careful with your words. You are now a Lieutenant Commander. What you say carries great authority. You won't serve our crew well by challenging the Captain's authority."

"You don't understand. She's not fit to command the *Vindictus* in battle!"

Figo replied, “Sir, who are you to judge Astraea’s fitness? She is Captain. That’s her rank and duty. Captain Astraea was given no choice but to command us in battle.”

“But, Mr. Figo, look at her! For our gods’ sake, the woman wears laced armorettes and high heeled boots on our bridge! Shall we follow the orders of a Captain attired in the latest wedding fashion? She isn’t even wearing a battle blade!”

Dr. Medens interrupted, “Mr. Negatio, the Captain left her wedding when duty called her to war. The lace she wears was intended for holy matrimony. When you married sir, did your wife wear army boots to the alter?”

“Listen to me,” argued Negatio. “That woman has poor judgment. She was going to marry a local politician, a common ballot chaser!”

Doctor Medens shook his head dismissively. “Lieutenant Commander Negatio, Miss Astraea was called to duty in her Templum. She stood before her Gods in the company of family, loved ones, and friends. There, our Captain answered a call to war. Many others did not. How many of us are aboard this ship only because we were already here?”

“I don’t understand her orders,” protested Negatio.

The Doctor answered, “Perhaps, if you donned lace and more fashionable footwear, you would be less confused?”

Negatio angrily warned, “I tell you, gentlemen. ‘Lieutenant’ Astraea will get us all killed!”

Mr. Figo replied, “you ask us to mutiny, sir. The answer is ‘no.”

Frustrated, Negatio pounded the table and declared. “I ask only for competent command!”

The Doctor concluded, “Gentlemen, this meeting is over.”

The gathering dispersed. Negatio remained alone. He sat quietly at the table with elbows bent and his forehead rested in his hands. He worried for himself, his crew, the *Vindictus*, and Gemina. Moments later he was hailed, “Commander Negatio, you are needed in damage control.” Without hesitation, Negatio did as commanded.

Captain Astraea returned to the ship's bridge. She studied the faces of her bridge crew. They were somber, focused, and determined. There were no signs of fear or doubt. Their determined eyes looked upon Astraea for command.

Many of them would die in the coming battle, maybe all. It was unlikely that Astraea's promise to return to Fidelis would be kept. She may not exchange vows with her groom, but the Captain would make a solemn vow to her crew.

"Lieutenant Carpo, patch me through to all decks."

"Ship-wide communication are open, Captain," announced Lieutenant Carpo. "The crew awaits your orders."

"This is Captain Astraea," she began. "We face a hostile enemy. As you know, the 'Visitors' demolished Outpost Semotus without provocation. They murdered thousands of men, women, and children. The enemy refused their surrender. The attackers destroyed many of our finest ships, who answered distress calls from the Outpost. Other vessels were badly mauled. Now, the monsters race towards our world to inflict more death and destruction. Our loved ones pray for us. They wish we were not in harm's way. Yet, we would choose no other place at this moment. The *Vindictus* is the most powerful warship ever built. We are her crew. Together, we shall stand and fight in Sector 12."

Astraea continued. "We all expected Captain Timens to lead us into battle, or Commander Fugiens. Unfortunately, the Commander perished in an accident earlier today. His launch was wrecked in a transport collision. Captain Timens was badly injured in the same incident. So, here we are. I stand on the bridge as your Captain. I ask you to have faith in me, as I have faith in each of you.'

"I tell you now," Astraea vowed. "We will defeat the enemy in Sector 12. We shall destroy them all. They will regret their actions this day. None shall survive to boast about what they did to Outpost Semotus or to our brothers and sisters in arms. Death awaits them all."

Astraea paused. Remembering her beloved Fidelis, she concluded, "Our faith in one another will lead us to victory. This I solemnly promise each of you."

Slowly, quietly, crew members on every deck began tapping their feet. Others joined in unison. They began stomping and chanted Astraea's name for twenty seconds. Afterwards, everyone went to their action stations.

Mr. Telum took command of the ship's weapons systems. He asked, "Your orders, Captain?"

"Take the *Vindictus* to the coordinates given, Mr. Telum. We will anchor ourselves to the planet Immanis."

No one understood the Captain's command. However, it didn't require their judgment. Astraea's orders would be followed. The *Vindictus* crew had faith in her and each other. That made them a powerful force. Their Captain promised to destroy the enemy, and the crew believed her. They were willing to follow Astraea through death's doorway.

The mighty *Vindictus* headed to War.

[End of Chapter]

CHAPTER EIGHT

Frugie's Fight

As the clouds of war approached Gemina, senior officers were called to duty. One was Commander Evanen Titos Scitus. He was Captain of the *GSS Tigress*, a powerful Class G Destroyer. She was the sistership of the *Lioness*. Both packed a powerful punch with six automatic pulse Cannons. The Commander's destroyer was fast, and she had an elite crew. She was a match for any warship.

Scitus was unaware that the Navy was searching for him. News of war had yet to arrive at Lucelenus Province. He was attending personal duties, most importantly his troubled marriage. His wife feared it was ending. She asked him to join her in marriage counseling. Neither of them knew that war would soon take him from her, perhaps forever.

Two marines were assigned to transport the Commander to the nearest Launch Pad. Corporals Probus and Excolo were directed to locate Commander Scitus and get him aboard a transport. If he resisted, they would place him in custody.

The marines first visited the Commander's home. No one answered. An elderly neighbor, Mrs. Comis, called to them from across the street. She informed the Corporals that the Commander and his wife were going to marriage counseling. To their surprise, she even gave them the counsellor's address. That was the Marines' next destination. They would find Scitus there.

The Commander's wife, Felicia Frugie Scitus, neared the marriage counselor's office. Frugie attracted attention as she hurried down a crowded sidewalk. Her trim, feminine figure and sweeping black hair pranced in lively harmony. With head held high, her dark, brown eyes searched ahead. She believed counseling was necessary to save her marriage.

Her journey ended at a sorrowful address, "Deep Six Discidium Plaza." Above its entrance was a sign that read, "Exspes Marriage Counseling."

Frugie looked about for her husband. He was nowhere in sight. After waiting several minutes outside for his arrival, she entered alone. A frumpy, irritated woman greeted her. "Good morning. Are you Mrs. Scitus?"

Frugie nodded affirmatively.

"Doctor Celsus is waiting. She wants to know if your husband is coming?"

Frugie took a seat in the waiting room. She set her handbag beside the couch. "He will be here," she answered.

"If the Commander comes," continued the receptionist impatiently, "the Doctor will see you both. Otherwise, she may offer you individual counseling."

Several minutes passed. There was no word from Scitus. The Counselor marched into the waiting room. She cast a disdainful look towards Frugie. "Mrs. Scitus, it doesn't look like your husband is coming. We will wait for him in my office. Follow me, please."

Frugie was disappointed that Scitus had not arrived. She gathered her handbag and left the waiting room. Celsus led Frugie into an impressive office. Its walls were paneled with walnut veneers, diplomas, and certificates. The Doctor sat behind a well-polished desk. Frugie saw her own reflection on its surface. Somehow, that was appropriate. She needed introspection to find answers to her marital issues.

Dr. Celsus was an elite member of the professional class. Her attitude neared condescension. The Doctor tried to obscure her own feminal beauty. She was blessed with the sculptured features of a fashion model, and long, gorgeous hair. However, Celsus needed her clients to see her as a studied counsellor, not another "pretty face." So, the counsellor wore her beautiful brunette locks in a bun, which gave her a serious, stern appearance. She hid behind a constant frown. Some called it her "bitch face." Surprisingly, many men found it attractive.

Frugie didn't care about the Doctor's posturing. The two measured one another across the walnut desktop. After evaluating her client, Celsus began a conversation.

"Would you prefer that I call you 'Frugie' or 'Mrs. Scitus?"

"Either is fine."

"Very well. Mrs. Scitus, do you really expect your husband will attend marriage counselling with you?"

"He will be here, soon."

As the Doctor spoke, her eyelids nearly closed. That presented her client with a bored, detached look. Celsus impatiently glanced at her time piece. "Don't be upset," she commiserated, "but many times, they don't come."

"My husband will be here," insisted Frugie.

"Mrs. Scitus, has your husband missed engagements with you before?"

"Yes," she answered.

Dr. Celsus impatiently studied her crowded appointment calendar. "Would you like to reschedule your session?"

"No, this one is not over."

Frugie's insistence that her husband would arrive struck Celsus as odd. Perhaps, it was a clue to her client's issues? "Does his absence upset you?" she asked.

"Yes, it does."

"Are you angry?"

"Yes, I'm angry. That's part of my problem, Doctor. Give him five more minutes, please."

"Very well. While we wait for the Commander, please tell me more about yourself."

"I'm a nurse at Campos Medical Center. Scitus and I have been married twenty-one years. We have two children, a daughter named Spes and a son named Versutacus. Spes is a student at Campos Provincial University. Our son is an Ensign in the Galactic Navy. He's a communications officer."

"I see from your intake that you're forty-one years old. You have a very youthful appearance, ma'am." Unconsciously, Celsus compared her own appearance to that of Mrs. Scitus, who impressed her. "Does it bother you that your son has followed his father's footsteps by joining the Navy?"

"Not at all. I'm proud of them both."

Celsus closed her notebook and laid her pen on the desk. She thumped her fingers atop Frugie's reflection on her desktop. The disappointed counselor was about to conclude their meeting when her receptionist interrupted, "The Commander is here."

That surprised the Doctor. She had awaited disappointment.

Celsus anticipated Frugie's husband would be a stiff, disciplined military type with anger issues. The doctor intended to examine and dissect Scitus in his wife's presence. Many wives enjoyed that. "Send him in, please."

There was a light tap on the door. The Counselor brusquely answered, "Come in." Her office door opened slowly. The Commander hesitantly entered. The Doctor carefully surveyed him. Scitus was a tall, handsome man in his early forties. Unlike his wife, he was tanned and had green eyes which were common in his Province. He kept his light brown hair trimmed in a crisp military cut. Dr. Celsus assumed Scitus wore his military uniform to assert supremacy.

The Commander was glad to see Frugie. He removed his hat slowly. Then, he unholstered his pistol and laid it on a chair. As with most officers in the Galactic Navy, he also carried a sheathed battle knife on his left thigh. Per military protocol, he didn't remove it. "Ladies, please accept my apologies for being late." His voice was firm, as one would expect from a Commander. Yet, it carried a tone of assurance and empathy.

The first sight of Scitus paused Celsus. He was an attractive man. The Commander fit his uniform well, and there was something about his weapons that were exciting. She noticed his large hands. They were appropriate for a man of his height. Celsus grew furious with herself. To her surprise, she was aroused! It was inappropriate and unprofessional. Yet, she couldn't control the Commander's unintended effects upon her.

The doctor lost her bitch face. She sought to regain her composure. She bit her lower lip and crossed her legs. Celsus averted the Commander's green-eyed gaze.

"Good morning, sir." She welcomed him. "We're glad you could make it."

"As am I, ma'am." The Commander sat beside his wife. He gently took her hand. He gazed into Frugie's eyes and smiled. She placed her other hand on his. They obviously were in love.

One of the Doctor's tactics was to challenge the offending spouse. She believed it exerted her professional authority and broke down his resistance. The Doctor also wanted to end her unwelcome attraction to Scitus. In a stern voice, Celsus scolded the Commander, "You know you're late, sir?"

"Yes, Ma'am. I apologize to you and Frugie."

"Are you upset by your wife's request for marriage counseling?"

"No, ma'am."

"Very well. Let's proceed." The Doctor asked Mrs. Scitus to explain why she arranged counseling. "Tell us why we're here?" she queried. "Has the Commander abused you? Has he struck you physically?"

Mrs. Scitus appeared surprised. "No, of course not!"

"Has he abused you emotionally?"

"What?"

"Has your husband ridiculed you? Has he insulted you, or embarrassed you before co-workers, friends or family?"

"No, Doctor!" Frugie vigorously shook her head negatively. She looked to her husband with surprise. That wasn't why she wanted counselling.

"Has he called you demeaning names?"

"Demeaning names?"

"Has your husband called you a pig, whore, bitch, or some other ugly name?"

The Commander recoiled from the Doctor's hostile remarks. He would never utter such ugly words to Frugie. If Frugie had not requested his presence, Scitus would have left. Frugie answered, "Never!"

The Doctor seemed surprised. She held a pen above her note pad. "Madam, for what issues have we brought Commander Scitus here?"

Frugie had difficulty answering. Her silence lasted moments. Meanwhile, the Counselor's mind drifted. She wondered whether she should have a private session with the Commander. Unconsciously, Celsus removed her hairpin. Her silky auburn hair fell below her shoulders and swayed as she turned her head. The Doctor was still aroused. That caused her to grow angrier with herself. Celsus was determined to regain self-control. "Mrs. Scitus," she asked, "why have you asked for marriage counseling? What fault has your husband?"

"It's not his fault. It's mine."

"Ma'am? Have you been unfaithful to your husband?"

"Never."

"Have you been abusive to him?"

"In a manner, I have."

"Have you struck him?"

"No, Doctor."

"Have you abused him emotionally?"

"I'm afraid so."

Dr. Celsus was puzzled. She observed Mrs. Scitus. What she saw was a conflicted woman who desperately wanted help. Frugie teared, as the Counselor pressed for clues. "Mrs. Scitus?" she asked. "Please explain your problem."

With watery eyes, Frugie looked toward her husband. She took his right hand and buried her cheek in his palm. She kissed his fingers.

Celsus asked again. "Tell me, Mrs. Scitus. What is your problem?"

"I grow angry at my husband."

"Why?"

"I love Scitus with all my heart. I love him more than anything. I need him to love me back. I want him to love me as I do him."

The Counselor paused. "What do you mean by that?"

"It hurts me when my husband answers another before me."

"But you said he was faithful?"

“That’s not what I meant. My complaint is that he answers the needs of others, before mine.”

“Who?”

“The Navy. He travels the globe when called. He even fights pirates in space! He does whatever the service asks. It doesn’t matter how I feel. I don’t understand why he leaves me alone. I feel abandoned, unloved. That hurts worse than anything!”

“How do you answer that, Commander?”

Scitus ignored the counselor. It was his wife’s words he heard. She was in pain. Never had he understood it.

“Please, look at me.” he asked Frugie. “I do love you more than anyone or anything.”

“How can you say that? Even when we first met, it was my dog you loved!”

“Your dog? You mean little Prowser?” Scitus smiled fondly at their pooch’s name.

Counselor Celsus watched intently and listened. Professionalism required it.

Frugie answered. "It was Prowser who caught your attention. That's why you spoke to me. You asked what breed he was? Then, you told me how Prowser was the happiest pup in the park."

Scitus searched into his wife's soulful eyes. He saw she didn't understand him. Before he could explain, there was a knock on the door. The receptionist called Dr. Celsus over the office-com. "I tried to stop them," she protested. The door opened. Two Marines entered. They were Corporals Probus and Excolo. "Commander Scitus?" one asked. It was obvious they found their man.

He answered, "Gentlemen, remove yourselves from this office immediately."

"We're sorry, sir. We can't do that."

Scitus rose to his feet. "I am ordering you, Corporals. Leave or I shall remove you."

Neither Marine expected a naval officer to threaten them with violence. That impressed them! However, they had to carry out their orders. "It's Navy orders, sir. You're needed immediately."

Frugie knew what that meant. Scitus would leave her again because the service needed him. She resented it. She always feared he wouldn't return.

One of the Marines told him, "We're at war, sir." The other drew handcuffs from his belt. "Sir, do we have to cuff you?"

Scitus was puzzled. "Why would that be needed?"

"Others haven't come willingly, Commander. Outpost Semotus was attacked. A large enemy fleet is quickly approaching. All ships have been ordered into battle You've been called to duty."

Frugie wept. "You'll go with them to war? You're going to leave me alone? Dear husband, it breaks my heart."

The Commander gently ran his fingers through her dark hair. He looked down upon his wife and replied.

"Yes Frugie, I will go to war, but you should know why."

His wife wept quietly.

One Marine stepped forward. "Commander, we must go now. Hades is consuming everything. The Navy needs you up there!"

With a slight nod of his head, the Commander acknowledged the Marines. He understood the urgency of war. He would fight that battle after saving Frugie from her own.

"Frugie, have you wondered why I was in the Park the day we met?"

Before she could answer, he continued. "I was lost. I was a young, conflicted ensign who had answered his call to duty. I devoted myself to the service. I didn't understand why it was my duty, instead of another's. I couldn't find an answer. Then, I saw you. It was as if clouds parted beneath the sun. No longer was there darkness. I saw my answer. I found my purpose.'

"Frugie, you gave life to the whole park. I wasn't the only one affected by your joy. Little Prowser yelped and ran in circles with his tail wagging. He wanted your attention. Children watched him and laughed. Couples smiled at one another in amusement. I remember how an older lady melted into her husband's arms as they sat on a bench watching the love around you. You were my answer. I fell in love with you then. I remember everything from that afternoon. I remember your voice and laughter. I remember how you wore your hair and the colors of your dress. Then, you saw me and smiled. I was lost in your brown eyes. I felt an unbreakable bond between us. How could you not have felt it, too?"

One of the Marines interrupted. "Sir, we must go."

Frugie asked the Marines, "Gentleman, spare us one last moment." Then, she turned to her husband, "I felt it," she answered. "I never knew you did. I'm sorry."

"Frugie, I love you more than anyone or anything. I willingly answer duty's call. I go to war for you."

One of the Marines approached the Commander with handcuffs. "Sir, we haven't any more time." His voice was impatient.

The Commander held Frugie closely and kissed her. When their lips parted, each knew it may be their last moment together. Before he left, Scitus made the promise which warriors never expect to keep. "Don't worry. I'll return soon."

Frugie and the Doctor were speechless. The Commander turned and told the Marines, “Gentlemen, we have a war to fight. Let’s go.” They left hurriedly. Counseling had saved Frugie’s marriage. What she couldn’t know was that it saved her husband’s life. If not for the slight delay caused by their session, he would have died in the same terrible launch accident that severely injured Captain Timens of the *Vindictus* and took the life of his First Officer.

There was silence in Celsus’s office. Moments later, it was interrupted by weeping. Both women sobbed. Perhaps it was unprofessional, but Dr. Celsus held Frugie’s hand. Her own cheeks were wet with tears. “Be strong,” she urged. “He loves you.”

For the first time, Frugie understood why.

[End of Chapter]

CHAPTER NINE

Journey Past Hades

Commander Scitus rode to the Launch Zone with the Marine Corporals. He ordered them to explain what was happening. Being typical Marines, they acted too stupid to understand the Commander. After all, he was Navy. So, naturally, their duty was to aggravate him.

Probus began first. He knew they wouldn't be court-martialed for irritating the Commander. They were at war. Death had visited Semotus. It was fast approaching Gemina. The Corporal knew his superiors weren't going to put Marines in the stockade. So, Probus decided to annoy the Navy Commander with marriage counseling jokes.

"Hey Commander," he called.

Scitus answered, "Yes, Marine."

"Commander, me and Corporal Excolo apologize for interrupting your marriage counseling."

"That's alright, Corporal Probus. Thank you for your patience."

"Sir, Corporal Excolo had marriage counseling."

"Is that so? I didn't know he was married."

"He wasn't sir. The Counseling worked."

Excolo chuckled.

"Is that right Corporal Excolo?" asked Scitus.

"Yes, sir."

Probus continued. "Would you like to hear about Excolo's counseling, Commander?"

Scitus knew he was being pranked. Other than fighting and cussing, that's what Marines do best. They still had ten minutes of travel, so he answered, "Sure, tell me about the Corporal's marriage counseling."

Probus continued. "Very well, sir. Excolo's therapist was real smart. She had a theory that couples who has sex each day is happiest. So, she decided to write a paper and become famous. For proof, his therapist invites all her clients to a seminar about sex. Of course, that included Corporal Excolo."

"Of course, Corporal."

"Nobody knew why they was called to the seminar. So, they was all standing there waiting for counseling when she asks, 'How many of you has sex once a day?' Half of everybody raised their hands. Each of them had a big grin. Then, she asks, 'how many has sex once a week?' A third of everybody raised their hands. They grinned a bit less. Then, the Counselor raises her eyebrows. She asks, 'Okay, how many has sex once a month?' Just a few hands went up. They didn't look happy at all! Then she asks, 'Okay, how many of you has sex once a year?' So, Corporal Excolo jumps up. He had the biggest grin and was happily waving both his hands! The therapist was shocked by his joyful display. She thought Excolo may be punking her theory? So, she asks the Corporal, 'If you only has sex one day a year, why is you so happy?"

"That's a good question," replied Commander Scitus. "How did Corporal Excolo answer her?"

"Corporal Excolo yelled, 'I'm happy, because today's the day!"

Excolo grinned at the joke. That mildly surprised the Commander.

"Corporal, why are you grinning?"

"Because that was last year. Tomorrow is my next day."

Probus laughed loudly.

Scitus shook his head dismissively with a pranked smile. "You know Corporals," he began. "I'm getting to like you, even if you are Marines."

Probus had more material. They were almost at their destination, and he didn't want to waste it. So, the Marine asked Commander Scitus if he would like to hear more about Excolo and his wife?

Scitus was surprised. "Corporal Probus, you told me Excolo wasn't married."

Probus shrugged his shoulders. "That was later," he said.

"Excolo" Scitus queried, "are you married?"

He answered, "Yes, sir. Me and my old lady got married. It will be a year from tomorrow."

"Alright, let's hear about the Corporal and his wife."

"Okay, this is the truth," began Corporal Probus. "On the first night of Excolo's honeymoon, he ain't sure how to tell his new bride about his stinky feet and smelly socks. What he don't know is she wonders how to tell him about her nasty breath! Until then, she was able to hide it. Anyway, Excolo got real brave. He says to her, "I has a confession."

So, his old lady draws closer. She peers in his eyes and says, "Corporal Excolo, I has a secret, too!"

"You has a confession?" Excolo asks his wife. "Don't tell me—you been eating my socks!"

"Since that day, Excolo is convinced his old lady's stinky breath was from eating his socks. She never said otherwise!"

The Commander chuckled. "Is that so, Corporal Excolo?"

He replied, "Aye, sir. The old lady got stinky breath!"

Probus grinned. "I love Corporal Excolo, sir. He's a good Marine." Then, the Corporal noticed something was wrong. The sky glowed ahead. It was ablaze. Emergency vehicles passed from every direction with screaming sirens. The Corporal's mood changed.

"Sir, it looks like something bad happened ahead," he told the Commander.

"We'll find out soon enough, Corporal."

Probus received a transmission. "Commander, they tell me Outpost Semotus was attacked. It's bad up there."

Scitus acknowledged the Corporal's remark. "Then, that's where I'm needed most, Corporal Probus."

"Commander?" queried Probus.

"Yes, Marine?"

"A lot of people already been killed, sir. Some say Semotus was wiped out. They say nobody survived. Whoever the enemy is, they're coming here. I hear they got a lot more ships than us."

"Our Navy may not be as big as theirs, Marine. But a Gunny Sergeant once told me, 'It's not the *size* of the dog in a fight that matters. It's the size of the fight in the dog." Scitus added, "Corporal, we have a lot of fight in us." After a pause, Commander Scitus asked for a promise. "If they get here, Corporals, promise me that you Marines will make them regret it."

"Aye, sir. If them things get past the Navy, we will kill them all, sir. Your wife will be safe."

The trio entered the Launching Area. As they did, several ambulances and fire trucks raced by them. Flames reached into the skies. Black smoke slithered between every structure. Corporal Probus stopped beside a busy military patrolman and asked, "Hey, Sarge, what happened?"

"There was a tragic accident. Two DUKs collided. One went out of control. It veered into a Launch and blew up. Lots of people died." Probus witnessed the blazing wreckage. Bodies sprawled across the site, some aflame. Med techs were giving aid to survivors. Ambulances were loaded with injured or dying personnel. Many stunned people wandered aimlessly. There was yelling, crying, and screaming. It was a nightmarish scene. Corporal Probus would never forget it. From that moment, he feared flying, especially aboard DUKs.

Scitus inquired. "Sergeant, my ship is the *Tigress.* Is our launch ready?"

"No, sir, Commander. Your ship launched without you." The Sergeant added, "The *Vindictus* ain't got a captain, either. Captain Timens was injured in the accident. His Exec was killed. Someone said *Vindictus* is going to war with a Lieutenant commanding her."

That was bad news. Warships, including his *Tigress*, needed veteran leadership. Yet, the planet's mightiest ship, the *Vindictus*, was commanded by a junior officer? It didn't bode well for Gemina.

"Corporal," he asked. "Do we still have Galactic Cutters?"

"Yes, sir. We have six in the South Hanger."

“Take me there.”

“Sir?”

“Take me to the cutters.”

“Commander, with all due respect. You can’t fight nobody in a Cutter! They’re fast, but they ain’t got much punch. All they got is small caliber automatic cannons. That’s it.”

“Gentlemen, don’t you go to battle with pulse rifles?”

“Aye, sir.”

“Do you kick ass?”

“Aye, sir.”

“Then, why should I do less?”

Corporals Probus and Excolo took Scitus to Hanger Three. Six cutters awaited inside. Their engines were powering. Several young ensigns busily inspected their ships. They were cutter pilots. Each wore flight gear and awaited orders that would never come.

Probus stopped the Corporals' vehicle. "Here you are Commander."

"Thanks for the ride, Marines. May the gods be with you."

"You, too, sir."

Probus and Excolo saluted the Commander. He returned their salutation. Then, they drove away. War was coming for them, too.

[End of Chapter]

CHAPTER TEN

Fourteen Minutes

Astraea was Captain of the *Vindictus*. Never had she commanded a man-of-war, nor had she led a vessel into combat. Within moments, she would command Gemina's only dreadnaught against a rampaging horde of malevolent warships. Trained at the Academy, Astraea understood that most battles are desperate affairs. This would be no exception.

Fleet engagements are often decided in minutes. Fourteen minutes would decide the coming conflict. With that time, Astraea and Fidelis could have exchanged their wedding vows. Juvo and Fidelis shared a bowl of stew. It takes longer than fourteen minutes to do a load of laundry.

The powerful *Vindictus* raced towards Sector 12. It was a "No-Go Zone." That designation meant it was too dangerous for navigation. Ships were forbidden to go there. However, young Astraea took the Navy's only dreadnaught and her crew into the dark, menacing expanse. Whatever hazards awaited them could be worse than the oncoming enemy armada.

The 'Visitors' pursued three wounded Geminum warships into Sector 18. An unrelenting battle approached. They had smashed Outpost Semotus and obliterated four of Gemina's most powerful warships, The fate of the *Vindictus*, her crew, and their world would be decided in a dark, forsaken void.

Captain Astraea understood the importance of time. She would use it as a weapon against the enemy. "Mr. Telum," the young, lissome Captain asked, "what time is it?"

He answered, "It's 10:36 a.m., Ma'am."

"What is our time of arrival, sir?"

"At maximum speed, we should reach Planet Immanis by 11:12 a.m."

"Mr. Telum, when will we rendezvous with the *Lioness* and her squadron?"

"If they survive, ma'am, the three ships will rendezvous with us at 11:14 a.m. The enemy won't be far behind."

"Continue our course, Mr. Telum."

During the following 36 minutes, the *Vindictus* raced to Immanis. As they slowed below FTL (Faster than Light Speed), a solar system came into view. They were in Sector 12. It was a Red Sector.

"We're here, Captain," acknowledged, Lieutenant Carpo.

Nothing about Sector 12 appeared unusual or hazardous. There was no unusual radiation, gravitational holes, or other threats. It appeared quite ordinary.

"Captain, we're in a moderate gravity current. It's flowing from the asteroid belt towards Immanis. I recommend we make course adjustments."

"Agreed. Mr. Telum, make it so." Gravity waves are common in deep space. Graviton particles often flow between objects with mass. Mr. Telum made a slight change of course.

"We're nearing Planet Immanis, Captain."

"Hold course, Mr. Telum."

It was 11:12 a.m. when Mr. Telum announced the ship had reached its destination. Ahead was a distant yellow sun. In its orbit was a shiny, reddish ring, which Captain Astraea recognized as the Mordices Asteroid Belt. Further from the sun, a gigantic, gaseous planet completed its solar orbit every six days. It was named "Immanis." The speed of its circumgyration prevented Immanis from rotating. Consequently, half the world suffered relentless sunshine, while the other remained in total darkness.

"Captain, the Asteroid Belt has an elliptical orbit around the sun. Its asteroids are nearing apogee. Ma'am, it's very unusual. There's an abnormal number of asteroids, rocks, ice, and debris in the belt. Its mass is enormous. Continent size chunks of broken planets are adrift in it. The kinetic energy there is beyond measurement. Anything drawn into the Mortices Belt will be crushed and pulverized. Ma'am, we must keep our distance."

Astraea acknowledged, "Understood, Mr. Telum."

"Captain, if it's the asteroid belt that makes Sector 12 a no-go zone, we can avoid it by maneuvering at a safe distance."

"Are there any other anomalies, Mr. Telum?"

"I can't say it's an anomaly, but Planet Immanis also has an eccentric elliptical orbit. The planet is nearing perigee with the Mortices Belt."

"Thank you, Mr. Telum."

"Captain," reported Miss Carpo. "Scans show three Geminum warships approaching rapidly from Sector 18. They are pursued by more than sixty enemy vessels."

"Miss Carpo, please advise Commander Clades of the graviton current. He can use it to evade the enemy."

"Aye, sir. Message communicated. Captain, we have sixty-seven warships approaching at FTL. They include our ships - the *Lioness*, *Inflexible*, and *Indomitable*. The others are hostile. They're exchanging weapons fire."

"Lieutenant Carpo, call the crew to battle stations."

Klaxons throughout the ship sounded. Miss Carpo announced over the intercom, "Battle Stations, All hands man your battle stations. This is not a drill. I repeat this is not a drill." Everyone knew they were headed into a desperate fight.

Lieutenant Carpo announced, "Captain Astraea, the Commander of the *Lioness* is hailing us."

"Put him on, Lieutenant."

"*Vindictus*, this is Commander Clades of the *Lioness*. We've suffered heavy damage. Our weapons systems are compromised. Our engines are leaking plasma into space. We're making repairs.'

The Commander continued, "The enemy's weaponry is unfamiliar. I recommend your ship change its shielding and weapons frequencies to the following." He then transmitted various readings. "We found those more effective than others."

"Make it so, Mr. Telum. Lieutenant Carpo, order all other ships to switch shielding and weapons frequencies as directed."

"Commander Clades, take your ships beyond this solar system. Do not return until the enemy is defeated. Repair your warships. Rendezvous with other Geminum vessels and regroup. You are Gemina's last line of defense."

Mr. Clades replied, "Captain Astraea, damage control parties estimate our repairs will be completed in ten to fifteen minutes. Several frigates are joining us."

"I understand, Commander, but do not return your ships to this solar system. Am I clear, sir?"

"Aye, ma'am," the Commander reluctantly replied.

Mr. Telum prepared to put the *Vindictus* into orbit. He asked Captain Astraea, "should we put her behind the dark side of the planet, Captain? The enemy may not detect us there."

"No, Mr. Telum, place the ship in orbit on the solar side. Let them see us."

"But, Ma'am, we will be an easy target!"

“I agree, sir. Please follow my command.” Mr. Telum was puzzled, but he placed the *Vindictus* in orbit on the planet’s side facing the sun and asteroid belt.

What time is it, Mr. Telum?”

“It’s 11:14 am, ma’am.”

“When will the enemy be in range, sir?”

“They will be in range of our pulse cannons in ninety-three seconds.”

“Engage the enemy then with all pulse batteries, maximum fire.”

“Aye, ma’am.”

When the enemy came within range, the *Vindictus* unleashed a furious volley of cannon fire. Three enemy ships were immediately struck and destroyed. There were cheers among the crew.

Two Gemina frigates followed the *Lioness.* The gravely wounded *Indefatigable* and *Indomitable,* limped past the *Vindictus*. Plasma bled from their ruptured hulls into the void.

Valiantly defending their withdrawal was the *Lioness*. She had suffered extensive wounds. Green plasma hemorrhaged from the Destroyer's stern, but she and her crew were unbowed. The *Lioness's* angry cannons defended her and the two frigates. They repelled enemy strafing runs.

"Commander Clades," instructed Captain Astraea, "withdraw your ship immediately."

"Negative, Captain Astraea."

"Commander Clades, I order you to withdraw the *Lioness*, now."

"Captain, should we leave, your flank will be unguarded."

"Withdraw, sir. Rendezvous with the remainder of the fleet. Repair your ships and regroup a squadron." Clades reluctantly obeyed Astraea's command. The veteran officer questioned the wisdom of the young Captain's order. However, he would not break the military chain of command during combat. He did as she commanded.

The *Vindictus* unleashed her powerful graviton cannons against the enemy. Four of their warships were obliterated. Several interceptors veered away from the *Lioness* to target the *Vindictus*. The dreadnaught protected the *Lioness* and her battered companions by unleashing withering cannonades against the enemy. That attracted the attention of dozens more enemy vessels. The "Visitors," as Prime Minister Morus had called them, began making dogged, sweeping attacks against the powerful Geminum battleship.

"Captain," Mr. Telum reported, "they're attacking us with crossing squadrons. It's like a pack of wolves attacking an elephant."

"They will be surprised to learn that this elephant has claws and fangs, Mr. Telum."

Lieutenant Carpo warned, "Captain we're being drawn into the planet's gravity well." That didn't surprise Astraea. She saw it as a tactical advantage. The attackers' firing solutions were compromised by the giant planet's gravitational force and its ionized atmosphere. The Visitors were unable to land hits on the *Vindictus*. The Captain's risky plan of placing the ship into orbit was yielding results, but for how long?

Another six enemy vessels were blasted apart by *Vindictus*' s blistering cannon fire. Fifty-one remained in the fight. Another effective volley of graviton cannons destroyed eight more. The enemy frantically tried to resolve their firing solutions. Forty-three remained, then forty-one. More enemy warships were demolished.

Mr. Telum advised his Captain. "Ma'am, we're being drawn deeper into the planet's gravity well. If we don't exit soon, there will be no way to escape it.

"I understand, Mr. Telum. We stay here and fight."

Lieutenant Carpo cautioned, "Captain, there are reports from Engineering that we're falling into a gravity vortex. Mr. Figo says it will be difficult to escape, even at maximum propulsion."

The attackers adjusted their firing solutions. Their shots struck the battling *Vindictus*. The solar side of the dreadnaught was raked by enemy fire. She rotated her damaged side away from the attack.

"Mr. Negatio, this is Captain Astraea. We have rotated our damaged hull. Effect repairs, while our other side absorbs enemy fire."

"Aye, Captain," answered her damage control officer.

Despite Astraea's earlier order, Commander Clades led his small, battered squadron towards the fray. Green plasma swirled behind all three ships.

Lieutenant Carpo reported, "Commander Clades reports that he's bringing three warships to our relief."

"Commander Clades," directed Captain Astraea, "this is the *Vindictus*. You will return your ships to the coordinates previously provided. Regroup there. Repair and await further orders."

Reluctantly, Mr. Clades complied.

The *Vindictus'* hull was raked by heavy cannon fire. Iridescent plasma streamed from her growing wounds into the planet's ionosphere. Glowing green streaks crossed swirling skies. It was an alarming sight for Commander Clades and his crew.

"She's badly wounded!" Clades lamented.

Ensign Carpo reported, "Commander, we've received reports from the *Inflexible* and *Indomitable.* Their repairs have been completed. Both are battle ready. They request permission to engage the enemy."

Lieutenant Paraclitis urged Mr. Clades to ignore Captain Astraea's order. "Sir, our crews want to fight. When we were ordered away from Semotus, twenty thousand people were murdered by the Prime Minister's 'Visitors!' The *Lioness* can make a difference here! We can flank the enemy while they've engaged *Vindictus.*"

Commander Clades sadly shook his head. "No, Leo. Captain Astraea ordered us to hold. We shall hold."

Paraclitis pleaded, "Commander, the *Vindictus* is taking a pounding. Captain Astraea lacks your experience. She was a Lieutenant a few hours ago. They say she was taken aboard the dreadnaught from her own wedding. The fleet needs your leadership. We can't stand by idly while our most powerful warship is being lost by a junior officer. Please sir, let us fight."

"We have a task, Mr. Paraclitis. Repair our ships, gather a squadron and ready a defense. When directed, we will engage the enemy."

Aboard the *Vindictus*, events worsened. "Captain, this is Engineering. We're losing propulsion. If we sink deeper into the gravity vortex, we may never get out."

"We stay, Mister Figo. The enemy is within six hours of our world. We stop them here and now. They will advance no further."

Lieutenant Commander Negatio reported. "Captain, this is damage control. Our condition is critical. We've lost three graviton cannons and seven pulse cannons. Propulsion systems have fallen forty two percent. We're adjusting shield frequencies, but the enemy is quickly matching our efforts. They're scoring hits. We have another problem. The planet's gravitational pull is stressing our hull. Unless we escape orbit soon, the *Vindictus* will be torn apart. We're running out of time!"

"Mr. Negatio, get those cannons back online." She continued, "We need more power."

"Ma'am, the ship has no more to give.!"

“We need more power, Mister. Find it.”

“Aye, ma’am.”

Commander Clades intently watched the desperate battle. His ship was sufficiently repaired to engage the enemy. The Commander wanted to lead his growing squadron into battle. However, Captain Astraea had ordered the *Lioness* to remain in place while the *Vindictus* suffered hostile retribution and growing casualties.

The dreadnaught fought ferociously. She was a glorious spectacle. Her skin was ripped and burned from relentless punishment. Bright green plasma trailed the *Vindictus* as she drifted deeper through the planet’s atmosphere toward an inescapable gravity vortex. The embattled dreadnaught and her crew pounded the enemy.

Only twelve enemy ships remained. Severely damaged, they withdrew in a disorderly fashion. They may have escaped the *Vindictus*, but they would not escape their fate. It awaited elsewhere.

“They’re retreating,” Mr. Telum announced. Some crew members cheered. Others were stunned, wounded. Many more had died at their action stations.

"Have we won?" asked Lieutenant Carpo.

Captain Astraea called Lieutenant Commander Negatio. "Damage Control, this is the Captain," she announced. "Damage Control?" No one answered. "Mr. Negatio, please report." There was no reply.

Mr. Telum notified Astraea. "Ma'am, damage control was hit. We don't know how badly."

The ship struggled to escape the enormous planet's gravity well. Grudgingly, the wounded dreadnaught began to free herself. As she struggled, Mr. Telum reported another sixty-four enemy craft were approaching Sector 12 at FTL. Silence overtook the ship. The *Vindictus* narrowly survived the first battle. Yet, another attack quickly approached.

The situation worsened. Mr. Telum announced yet another sixty-four hostile vessels were approaching, then another sixty-four. He added, "Captain, there's a convoy of seventy-two ships behind them. I count two-hundred sixty-four enemy warships. They include troop carriers and support vessels."

It was an overwhelmingly powerful invasion force. Between them and Gemina was a badly mauled dreadnaught and a small squadron of seven warships, three of which carried battle damage.

From Med Bay, Dr. Medens reported, "Captain Astraea, we have no more room for injured crew members."

Aboard the *Lioness*, Lieutenant Commander Clades evaluated the situation. Mr. Paraclitis urged him, "Sir, the *Vindictus* won't survive another attack, not against that many warships. She's sinking deeper into the gravity well. It's going to rip her apart if the enemy doesn't. We've got to pull her out and withdraw."

Clades agreed, but he wouldn't disobey Captain Astraea's orders. "Leo, we have our orders. Assemble the squadron." The Commander readied his ships for a counterattack once the *Vindictus* was lost. Clades would hide three fast frigates behind the shadowed side of Planet Immanis. His remaining vessels would draw the invaders away. That would allow the hidden ships to strafe the enemy's troop transports from behind. Disabling the transports might spare Gemina from an invasion. The same three frigates would then strafe support vessels and target their propulsion systems. Clades intended to cause enough harm to the vast armada to convince the attackers to withdraw.

Aboard the *Vindictus*, Mr. Figo pleaded, "Captain, the gravity well is pulling us apart. We've got to get out of here! We haven't much time."

Astraea asked her first officer, "What time is it, Mister Telum?"

"It's 11:43 ma'am."

"We shall remain in place," Astraea ordered. "Lieutenant Carpo, patch me through to the crew."

Ship wide communications opened. Astraea spoke to her surviving crewmen. "This is Captain Astraea. We're badly damaged. Most of you are injured, many severely. Crewmates have perished. We don't know how many, but more will follow. The *Vindictus* is trapped in a gravity well. More than three hundred enemy warships are approaching us at flank speed. They include troop transports to invade Gemina. Our situation is desperate. However, the *Vindictus* was built for desperate situations. You were trained to fight them. What we do today will be remembered by our world forever. The enemy shall also remember it. They will forever regret challenging the *Vindictus* and her crew."

Captain Astraea paused. "Know this. We will destroy the enemy, all of them. They will never reach Gemina. They will die here in Sector 12."

Mr. Telum informed the Captain, "Ma'am, it's 11:46 a.m."

Astraea continued, "I ask you to fight for fourteen more minutes. Give me that, and we will kill them all. I give my solemn word." Astraea ended her broadcast. The ship was badly mauled. Her crew suffered heavy casualties. Yet, they would give Astraea whatever she asked, so long as they lived.

"Mr. Figo, I need fourteen more minutes from Engineering."

"Aye, ma'am."

The enemy fleet quickly swarmed though Sector 12. They ignored Commander Clades' small squadron. Instead, the hostiles gleefully resumed vicious attacks against the

Vindictus. They knew that the dreadnaught was badly damaged by the first wave of "Visitors." The newer ones intended to thoroughly punish the dying ship. They unleashed their wrath upon her.

Initially, the second wave of attackers suffered the same problems as their predecessors had. Their firing solutions were inaccurate. The planet's gravity and its ionized atmosphere affected their weapons. They couldn't land shots on their target.

Although wounded, the *Vindictus* quickly obliterated twelve attackers, then another twelve. The *Vindictus* had destroyed a total of seventy-four enemy warships, not including the twelve that escaped with mortal wounds. Over two hundred more approached with weapons ports open. They likely had adjusted their weaponry. Soon, the badly wounded dreadnaught would feel their wrath.

Dr. Medens called the Bridge. "Captain, our sick bay is filled with wounded and dying crewmen. We've got patients lying in the corridor waiting for treatment. We're doing our best down here. If you could, Ma'am, please beat hell out of the enemy. Don't let us die in vain."

"Understood, Doctor. Please carry on."

The *Vindictus* unleashed devastating cannon volleys against the enemy. Many more attackers quickly disintegrated from her cannonades. Yet, there were still over two hundred more. The Dreadnaught sank deeper into the planet's gravity well. Even with full power, she couldn't escape. Her crew knew it.

The *Vindictus* was obviously doomed. She was being crushed by an increasing gravity vortex. Her attackers knew the *Vindictus* would soon implode. They could have left the dying ship and her crew to a certain fate. However, the Visitors were excited by their adversary's helplessness. It gave them great pleasure to reduce the mangled dreadnaught to a shattered, broken hulk. They gleefully used her for target practice.

The Visitors should have attacked and overwhelmed the *Lioness* and her small squadron. That would eliminate any resistance to their attack on Gemina. Instead, they delighted themselves by savaging a broken, dying *Vindictus* while the small Geminum squadron awaited their turn to fight. The enemy's powerful battle cruisers and dreadnaughts closed their distance with the badly mauled *Vindictus.* Their heavy weaponry was overwhelming. Swarms of interceptors raked the dying ship. She had no escape. Without doubt, the ship's fate was certain.

Mr. Telum suffered radiation and electrical burns. However, he continued his duties without complaint. The Executive Officer reported, "Captain, we've lost propulsion systems."

"It's not over yet, Mr. Telum."

The dreadnaught then took a direct hit. There was a tremendous explosion in the ship's stern. A large section of the dreadnaught's hull breached. Remaining systems began shutting down.

Commander Clades readied his squadron. They were determined to visit retribution on the alien attackers. The *Indomitable* and *Inflexible* reported they were battle-ready.

Many of the *Vindictus* bridge crew were wounded. Some were mortally injured. Captain Astraea was among them. Yet, she maintained command. Her officers continued their duties.

“Mr. Telum,” she asked. “What time is it?”

“Ma’am?”

“What time is it, sir?”

“It’s 12:00 p.m., Captain.”

Astraea nodded. She turned to the Lieutenant Commander and told him, “It’s time. The crew gave us fourteen minutes, Mr. Telum. Increase power to the ship’s dampening systems immediately.”

“We only have emergency power, Captain.”

“Use it.”

Commander Telum didn’t understand Astraea’s Order. However, it was too late to second-guess her. In any event, he hadn’t lost faith in his Captain. “Aye, ma’am.” His right hand was broken, so Mr. Telum worked with his bloodied left hand. “Emergency Power has been transferred to dampening systems,” he reported. Momentarily that transfer would prevent the ship from being shaken apart.

Astraea announced over the intercom to her crew, “This is Captain Astraea. I asked you for fourteen minutes.

I promised we would save our world. The Visitors will perish. Their warships will be destroyed. You have given every second I requested. Know this. Because of your dedication and sacrifice, Gemina has been saved. The enemy die here. Their defeat will be total."

"Do you give the self-destruct order, Captain?"

"Negative, Mr. Telum. Keep faith, sir." Before Astraea could speak further, she was overcome by her wounds. The young Captain's head dropped. Her body slumped into the Command Chair. Grasped in Astraea's left hand was her betrothal pin. Earlier that same day, it carried promises of love and happiness. Perhaps, in another life they would come true.

Overloaded and broken circuits throughout the bridge crackled, flashed and flickered erratically. They filled the bridge with a burnt, overwhelming, chlorine-like odor. Breathing was increasingly difficult.

"Captain," Lieutenant Carpo shouted without response. "Captain, wake up!" Tears streaked across the Communication Officer's cheeks as she looked around the bridge at her fellow officers. Most had succumbed to their injuries. The Ensign's duty was to transmit the Captain's last orders. "This is the *GSS Vindictus* to Lieutenant Commander Clades. By order of Captain Astraea, your rank is increased to Commander. You are to take command of the remaining Geminum warships. Withdraw them further from this solar system for the next two minutes. If the enemy are not then destroyed, organize a defense of Gemina. Good luck, sir." With that communication, the dreadnaught's fight had ended. Miss Carpo gasped for breath several times before she was overcome by toxic fumes.

The Visitors had visited great death and destruction upon the dreadnaught and her crew. However, in their lust for violence, they didn't realize their terrible mistake. Astraea promised her crew the attackers would all die. With that promise, their fate had been written. The *Vindictus* would not die alone.

From a distance, Commander Clades lamented the *Vindictus' s* fate. "Now, we shall watch a great warship die." He called upon his crew to match her valor. "We shall follow their example. The enemy shall not pass us without great suffering." The few remaining lights aboard the *Vindictus* flickered before she went dark. Enemy craft strafed her relentlessly. Only one cannon aboard the battered warship returned fire. It was manned by Corporal Iratus. With each pass, at least one attacker fell victim to his unerring aim. Eventually, the Corporal's weapon was silenced.

Following Captain Astraea's last words came a great disturbance in space. Scientists called it a "quantum tide." Sector 12 had been declared a "No-Go Zone," because of the quantum tide's threat to navigation. It was a powerful torrent of gravity currents and undercurrents between the Mortices Asteroid Belt and the giant planet, Immanis. As a Science Officer, Astraea had observed and studied the powerful anomaly. She used it as a weapon against the Visitors. They had yet to realize it.

At 12:00 p.m., Immanis' s orbit reached its maximum perigee with the Mortices Asteroid Belt. That caused a sudden gravitational increase which pulled loose a huge ocean of graviton particles that rushed forth from the Belt towards Immanis. Tremendous gravity currents carried swirling waves of photon particles, which gave a fiery appearance as they overtook "Visitor" warships.

The Visitors panicked. Before they could react, the quantum tide carried the entire Phasmii fleet towards Immanis. As it neared that planet, the tide's violent undercurrent yanked them back into the asteroid belt where the warships were shredded, crushed, and pulverized by megatons of rocks, ice, and stones. What few remnants that remained of the enemy vessels were rudely spewed back towards Immanis, where the tide's undertow again swept them back into the asteroid belt for further shredding. Before the ebbing tide finally subsided, nothing remained of the 'Visitors' and their warships, except for scattered, subatomic particles. Their attack ended in total defeat. As Astraea had promised her crew, the enemy were all dead. Their fleet was annihilated. Gemina was saved.

The great Tide's undercurrent pulled the *Vindictus* free of the giant planet's gravity-well. A powerful graviton current violently shook the ship loose before rapidly receding into the asteroid belt. The mangled dreadnaught drifted aimlessly in orbit.

Commander Clades was stunned by the astrophysical phenomenon. The battle had ended. The *Vindictus'* broken hull was freed from the planet's gravity vortex. “Look at her, Leo,” Clades told his First Officer. “Even in death, the mighty *Vindictus* is magnificent.”

A trail of glowing, green plasma bled from the dreadnaught’s torn body into the planet’s atmosphere. Her once beautiful white and blue hull was blistered, scorched, and breached. Debris emptied from her gaping wounds, including the remains of crewmembers who had perished in battle.

“This was her final fight,” declared Clades. “From this day forwards, the *Vindictus* shall forever be more than a warship. She shall be a legend.”

[End of Chapter]

CHAPTER ELEVEN

Doctor Varus's Departure

The Battle of Sector 12 had ended. As Captain Astraea promised her crew, the Visitors were annihilated. Their warships had been destroyed. The *Vindictus* had fought valiantly, but the mighty dreadnaught had been reduced to a dark, broken hulk. It was battered and appeared lifeless. Yet, there was hope that there may be survivors.

Aboard the *Lioness*, Commander Clades called a meeting of department heads to organize rescue teams. Their last meeting in Sector 16 had been contentious. Bitter feelings remained, especially towards Dr. Varus.

Ensign Sonos sat alone in her quarters. The Commander had not summoned her. In the past twenty-four hours, the young Ensign had lost her family and friends. She had no one, anywhere. Even tears abandoned her. Darkness was the Ensign's only companion.

Before her shift ended, Miss Sonos electronically transmitted court documents to Doctor Varus. It was her duty as the ship's communications officer. Sonos didn't know they included a decree that terminated the Doctor's parental rights. They also included a restraining order to keep him from going about or communicating with his children, who had been adopted by his ex-wife's new husband. He received the unsettling news after transferring seriously wounded crewmen to the *GSS Clementia*. Two died enroute.

Doctor Varus retired to his quarters. Awaiting him was the court order. He read and deleted it. As he tried to cope with the court decree Dr. Varus was beckoned over the ship's com. "Dr. Varus, please report to the ready room." Physically and emotionally exhausted, the doctor peered into his dresser mirror. As he leaned forwards, an exhausted, haggard man stared back in dismay. It was his own tortured soul.

The distraught physician opened a dresser drawer and withdrew a rapid injector. It was an advanced syringe used for medicating and vaccinating patients. However, he intended another use. The injector contained a lethal dose of relajante, a powerful muscle relaxer. The medication was designed to reduce pain. However, Doctor Varus would inject it into his forearm to end his own suffering. Its contents wouldn't mend his broken heart. Instead, they would slow it to a stop. The Doctor would die by relaxation.

As he raised the device, klaxons sounded throughout the ship. Varus tried to ignore the alarms, but they grew louder, as if the *Lioness* was screaming for him to stop. He closed his eyes and covered his ears, but her wailing continued. Reluctantly, Varus heeded the ship's calls. The *Lioness* needed her doctor. Members of his medical team rushed through the corridor outside his quarters. They answered the ship's alarms, as did Varus. He answered his call to duty. With a deep breath, the Doctor tossed the injector into a disposal unit. He hurried to the Ready Room.

On his way, Doctor Varus stopped outside a cabin. It was Miss Sonos' quarters. He knocked thrice against her door without reply. Inside, the young Ensign was oblivious to the wailing klaxons and commotion outside her door. She was lost in grief, but heard the Doctor's call, "Miss Sonos, are you there? May I speak with you." When she heard his voice, the Ensign assumed it was about the legal papers she had earlier transmitted. There was pain in the Doctor's voice, much as in her wounded soul. Reluctantly, she answered, "Enter, Doctor."

Varus had once been popular with the crew. That ended when he argued to Commander Clades that the *Lioness* shouldn't defend Outpost Semotus against the "Visitors." His arguments were personal, unpopular, and hurtful. Some believed Varus was a coward. In hindsight, his advice may have been correct. The Commander's plan to attack the Visitors would not have saved Semotus. The *Vindictus* later demolished all sixty-four of those warships. Yet, another three-hundred twenty-eight followed. Nothing could have defeated the Visitors, except the great quantum tide.

Dr. Varus entered the Ensign's disheveled quarters. She had been crying. "Miss Sonos, I hesitated to call on you, but the Commander is organizing rescue efforts to the *Vindictus*. The mission will be hazardous. I wanted to apologize to you before I left. I'm very sorry for your losses. I apologize for my hurtful words. They were inappropriate, unnecessary, and wrong. I should never have argued against fighting for your sister, aunt, and Outpost Semotus. I ask for your forgiveness."

The Ensign struggled to answer. Her nose was running, and her throat was sore from sobbing. Varus sat beside Sonos on her bunk and handed her his handkerchief. "I'm sorry, Miss Sonos. I shouldn't have imposed on you. I will leave."

As he stood and turned towards her door, she spoke. "Is that why you called? Doctor?"

"It is. I didn't mean to intrude."

"Doctor, I accept your apology. Whatever forgiveness I have is yours."

Her somber tone struck a chord of redemption within him. With pooling eyes, he answered with a broken voice, "Thank you, Miss Sonos. That means much to me."

"Doctor, wait for me. I'll go with you."

"I can't allow that, Miss Sonos. I've already disqualified you from joining any rescue team. Report to sick bay until this is over. Is that understood?"

"But, Doctor," she began to reply.

"That's an order, Miss Sonos." As he left her quarters, she answered, "Aye, sir."

Dr. Varus hurried towards the Operations Room. As he turned into another corridor, the Doctor noticed an open door. It was to Lieutenant Commander Paraclitis' s cabin. Varus stopped to accompany the young Navigator. However, before he rapped on the door, the Doctor heard a woman's voice inside. When he looked inside the cabin, there was no one there. What he heard was a vidcom which was looped to play again and again. He watched it twice.

The speaker in the video was an attractive, happy young woman. She had long, dark hair, which was swept over her left shoulder. Dangling from her right ear was an elegant, Adriana-blue topaz earring. It was encircled by bright gold vitaelium over a larger, two carat stone. Common to the Adrian Outpost, the stones matched the smiling beauty's sparkling blue eyes. In the background was a lush green garden, with colorful flowers that swayed gently in the breeze.

"My dear Titus," she smiled. "Your parents took me to dinner last night. My little brother, Traian, joined us. He's very fond of you. Now, he wants to grow up to be a

Stellari, and he's wearing navy blue to school. Your folks have been telling me old stories about you. Your sister and brother laughed, grinned, and smiled. Iris told me about the time when you found a toad she put in your bed. You may not think so, but it was very funny! We had a good time. Your folks know how you feel about me. They want me to love you, too. I didn't need convincing.'

"I've loved you from the first moment we met. I remember hearing your voice behind me. 'Excuse me Miss, is breakfast still being served?' I heard that same question a thousand times before, but your voice connected with me. I turned and saw this handsome Stellari holding a yellow lily. There was a sparkle in your eyes. At first, I didn't know whether to be upset, but your smile put me at ease. 'My name is Titus,' you told me. I accepted your flower and have fallen more in love with you every evening. Since you left, I've searched the skies for your ship and found it behind every star."

The young woman sampled the flower's fragrance. "I received your proposal. Your family knows about it. They're happy. They're going to meet you when the *Lioness* ports after your star-charting mission. They invited me and

Traian to go with them. I think they want to witness our engagement. We'll arrive a day early, so don't be surprised. I love you, Compertus."

After her cheerful smile, the telecom looped and began again. The Doctor was alarmed. He rushed from Mr. Hippolos' quarters and slammed shut the door behind him. The looped video continued playing. He hurried to the ready room. He hoped to find Lieutenant Hippolos there.

As he neared the meeting, the Doctor overheard Commander Clades ask, "Has anyone seen Dr. Varus?" No one answered. The Lieutenants were irritated by the mention of the Doctor's name.

"I'm here," Varus announced as he entered. Lieutenants Arma, Sarcio, and Hippolos ignored him. Mr. Paraclitis was on the ship's bridge, so only five officers attended. None of the Lieutenants welcomed the Doctor's presence. They remembered his cruel and insulting remarks at their last meeting.

Commander Clades hesitated to order rescue missions to the *Vindictus*. He and his crew had witnessed the violent quantum tide as it churned the entire Visitor fleet into space chum. If it returned, rescuers would suffer the same fate. Was the risk of losing dozens of Stellarii justified if there were no survivors to rescue? The great ship was dark, without power. There were no apparent signs of life aboard her. The dreadnaught was terribly damaged. Her structural integrity was questionable. She was about to fall apart.

"Gentlemen," Clades asked his officers, "are there any survivors to rescue?"

Sarcio answered, "Commander, our sensors haven't located any. However, their readings are unreliable. Our equipment is affected by damage aboard the *Vindictus*, including multiple disruptions from broken circuits and malfunctioning equipment. Damaged graviton and pulse weapons also skew our reception. We won't know about survivors until we get inside the ship."

Clades worried about the *Vindictus' s* condition. He asked Mr. Sarcio, "As the Ship's Engineer, how much time do we have before the *Vindictus* loses structural integrity?"

"She won't last long without being mended. We have no time to waste. If the ship breaks apart, survivors will quickly die of space exposure. Any rescuers aboard would perish, too."

"I see," Commander Clades acknowledged. "Only volunteers will be allowed on that mission."

"Lieutenant Sarcio, in your opinion, can we restore the dreadnaught's structure enough to tow her? If possible, we should take her home. We should return her fallen crew to their families and loved ones."

"We can't restore her life-support, propulsion or control systems, Commander, but I think we can make her strong enough to tow."

Doctor Varus addressed another concern. "Commander Clades, even if the ship's structure is repaired, it won't survive another quantum tide." We wouldn't survive it, either.

"That's a legitimate concern, Doctor. Mr. Paraclitis, when can we expect the next quantum tide?"

To that point, Lieutenant Commander Paraclitis had been silent. He stared at the floor, lost in thought. He raised his head and looked aside as he made calculations.

Paraclitis answered. He spoke slowly and quietly in a detached manner. “I can’t say for certain, Commander. However, my experience in Gemina's blue water Navy suggests the tide is cyclical. Gravity and water are much alike. They are fluid in motion and cohesive in nature. The quantum tide and its graviton currents behaved much like our ocean currents back home, such as in the Uldolmok Strait. They say it has “raging currents.” Its waters reverse directions every three hours. Ships were torn apart and sunk. If the quantum tide has such a cycle, we would have as much time as the *Vindictus* when she entered the solar system. We should have at least forty-eight minutes, if not longer. I can’t guarantee anything.”

“That may be enough time, gentlemen,” replied Clades. “We’ll have to hurry. This is an urgent matter.”

The Doctor stepped forward. He addressed Commander Clades, “I have concerns about certain officers.”

There were groans among the Lieutenants.

“Please, explain, Doctor.”

"I have medically disqualified Lieutenant Vox Sonos from any mission to the *Vindictus*."

Mr. Arma angrily retorted, "Varus is doing it again. He's trying to undermine your command, sir. The Doctor tried to disqualify Miss Sonos before the Visitors attacked Outpost Semotus. He discredited her for having family there. He also tried to disqualify Mr. Paraclitis, who had precious memories of the Outpost. There's nothing wrong with Ensign Sonos. We need our communications officer."

Before, the Commander replied, Doctor Varus added, "I'm also disqualifying Lieutenant Commander Titus Leo Paraclitis."

Mr. Sarcio protested. "Commander, I agree with Mr. Arma. Doctor Varus didn't like what Mr. Paraclitis said at our last officer's meeting. Now he wants to disqualify him. I'm sorry about your family issues, Doctor. You have a sad, miserable life, but that's no reason to disparage the crew."

Dr. Varus was hurt by his fellow officers' remarks. However, he answered, "What you say is true." He apologized. "Gentlemen, my remarks at our last meeting were intemperate, unfair, and cruel. There was no excuse for what I said. I apologize to each of you. I ask for your forgiveness."

Sarcio retorted, "We can't forgive you while you're trying to disqualify Miss Sonos and Mr. Paraclitis."

Varus addressed the Lieutenant Commander. "Please accept my apology, sir, but in my medical opinion, they're disqualified from any rescue mission to the *Vindictus*. That's my decision as ship's physician."

"Dr. Varus, what is the basis for your decision?" asked Commander Clades.

"On my way to this meeting," Varus answered. "I passed the Lieutenant Commander's quarters. His cabin door was open, so I looked inside to see if he would accompany me. I heard a woman's voice, but no one was there. She spoke on a looping vidcom in his quarters. In it was this remarkable young lady, who was very much in love with Mr. Paraclitis. She said she was going to travel to the ship's next port to accept his wedding proposal. Her little brother and the Lieutenant Commander's family were accompanying her.'

"So, what, Doctor," replied Mr. Arma angrily. "I saw the same video on my way here. Paraclitis is in love with a fantastic, beautiful woman, and she's in love with him. He's going to propose to her in the presence of his family, and she's going to accept. You may think that unfair because of your miserable life, but it doesn't disqualify him."

Dr. Varus paused before he asked, "Mr. Paraclitis, do you intend to return to your fiancée and your family?"

The Lieutenant Commander looked up for the first time in the meeting. He answered solemnly, "Yes."

"Mr. Arma," Varus inquired, "did you notice the time stamp on the vidcom?"

"I didn't have to. She said they were meeting Compertus at our next port."

"Mr. Arma, the vidcom was made two days ago. With that time and date, our next port would have been Outpost Semotus. That young lady, her little brother, and Mr. Paraclitis' family were all on Semotus when the Visitors attacked. All of them died. He now vows to join them."

Mr. Paraclitis dropped his head and turned away from the others.

Varus concluded, "I'm concerned for the well-being of Lieutenant Commander Paraclitis and Ensign Sonos."

Neither Mr. Arma nor Sarcio replied.

The Doctor continued, "Sending either into that wreck is the worst thing you could do to them. In its current condition, we may as well refer to the *GSS Vindictus* as the "*GSS Detritus.*" Inside her shattered hull are blood stains, gore, mutilated bodies and body parts. Sonos and Hippolos would see their loved ones' faces on every lifeless body within her. It would traumatize them forever."

Commander Clades turned to his Lieutenants. "Gentlemen, do any of you wish to be heard?"

No one replied. Mr. Sarcio carefully wrapped his arm around Mr. Paraclitis and led his shaken friend to Sick Bay. "I'm sorry, buddy," he said sadly.

Dr. Varus asked, "May I have permission to leave, Commander?"

"May I ask where you're going, Doctor?"

"With your permission, I will lead the ship's medical team to the *Vindictus*."

"Permission granted, Doctor. Good luck, sir. I wish you a safe return."

Minutes later, Dr. Varus led a determined medical team into the shattered dreadnaught.

He never returned to the *Lioness*.

[End of Chapter]

CHAPTER TWELVE

Funeral Barge

News of the battle was broadcast to Gemina. Juvo hurried to the Templum. He parked near the entrance and rushed inside. Kneeling at the altar was Fidelis. "I knew you would be here, Fidey," called Juvo.

"I didn't know where else to go."

Juvo gripped his friend's shoulder. "It's on the news. The battle is over. It's ended."

"Astraea, is she alright?"

"They said we won! The *Vindictus* destroyed hundreds of enemy ships. It's over!"

"Have you heard anything about Astraea?"

"Not, yet. Turn on your telecom, Fidey."

"I gave it to Puer this morning."

"Oh, yeah. We can use mine." Juvo pulled his telecom from his pants pocket. He powered it. There was a three-dimensional comview of the *Vindictus.* It was stock footage, showing a fresh, imposing Dreadnaught, ready for action. A proud, confident, smiling crew stood before her. She was an impressive sight.

The broadcast began. "Good afternoon, this is Campos News Channel Nine. The "Guests" invasion has been thwarted. Our Deep Galactic Fleet engaged hundreds of powerful enemy warships in Sector 12. Our defense was led by the *GSS Vindictus.* The powerful dreadnaught single-handedly destroyed over three hundred enemy warships." Cheers were heard in the background, but none from the *Vindictus.*

"What about Astraea?" queried Fidelis.

The comview showed the *Lioness, Indefatigable* and *Indomitable.* They had been mauled. The newscaster spoke. "Victory came at a price. We lost seven vessels. Three others were severely damaged. Even the mighty *Vindictus* took a beating." The comview switched to live footage of the dreadnaught. The ship drifted lifelessly in space. She had no spark of power or propulsion. The dreadnaught was a mere shadow of the powerful warship pictured in the stock comview. She was a ghost.

"Oh, my gods!" cried Fidelis, "Look at Astraea's ship! They killed her! They killed them all!"

"Hold on, Fidey. Nobody said that, yet."

"Look at the *Vindictus.* No one could have survived that!"

The newscaster continued. "Rescue crews have entered the *Vindictus.* They're led by Commander Clades. We have live comviews of their search. Stand by."

The picture changed. It showed hand lights flashing through dark, empty corridors. Hours before, those same corridors bustled with an active, lively crew. Now, there was no one.

Clades asked, "Have you checked Med Bay for wounded?"

"Sir, we checked. There is no Med Bay. It's gone."

"Give me a status report."

"It's bad, Commander. We're still searching. It's cold. There isn't much air." After a pause, "Sir, we found survivors in a few cannon pods." There were cheers in the newsroom.

Petty Officer Iratus stepped into the video. He had numerous abrasions and burns. Blood dripped from his nose. "There are survivors on Deck Two," he reported, "Keep searching."

"She's okay, Fidey!" Juvo exclaimed. "I know it. I just know it."

The newscast continued. It was cast holographically from Juvo's telecom across the room. "We're nearing the bridge. We found the ship's Engineer and several others below deck. They were sealing a hull rupture. We relieved them." The view then went blank.

"Juvo took his friend's hand. It's going to be okay, Fidey. I promise." Then, he bowed his head and prayed silently.

The telecom projected a new comview from a member of the rescue party. He said, "We can't get in the bridge. The doors are jammed."

"Break through them," ordered Commander Clades.

The vidpic lunged into the bridge as rescuers forced open its hatches. Lights searched the darkness for crewmen. "They're all down," announced one rescuer.

Clades ordered, "Check for vitals."

One searcher commented, "They took a terrible beating here, sir."

Broken monitors and other equipment flickered off and on. A damage alert whined through the ship's Com until a rescue party member turned it off. The sight frightened Juvo. He feared Fidelis would see Astraea there, dead.

"Don't look, Fidey. Please don't look." Juvo tried to shut off his telecom. Fidelis grabbed it from him. "I need to know, Juvo. I have to know. I'm watching. You can't stop me."

Juvo understood. He wrapped his arm around his mourning friend. He wouldn't let Fidelis collapse at the sight of Astraea's body.

The projected comview continued. "There are several survivors," reported a rescuer. "Some are in critical condition. All suffered radiation burns and exposure to toxic fumes. Many were killed in action, very many."

A member of the boarding party called out. "Commander, we just found the Weapons Officer, Lieutenant Commander Telum. He's still at his pod. "Mr. Telum is holding his weapon controls." The comview showed Telum sitting in his weapons control pod. "Mr. Telum?" one rescuer inquired. The Lieutenant Commander remained at his post without answering. There was a pause, before the rescuer announced, "Sir, Lieutenant Commander Telum is dead."

Juvo began shaking.

As the comview continued, lights searched through the darkness until they found another figure. "Sir, we've found Captain Astraea."

Fidelis stepped into the three-dimensional comview projection. He approached Astraea's image. He wanted to touch her, speak to her, but it was only a hologram. His hand swept through it. Astraea was motionless. She slumped to one side of the Captain's seat.

Juvo regretted viewing the news report. "Don't look Fidey," he pleaded. "They don't know what they're saying." He grabbed Fidelis by the arm and tried unsuccessfully to pull him away.

Fidelis couldn't believe what he was hearing and seeing. "They don't know what they're talking about," he said. "Astraea isn't Captain of the *Vindictus.* She's a *Lieutenant*! That can't be her. They have someone else confused with Astraea!"

A rescuer raised his voice. "She isn't moving." He called across the bridge, "Do you hear me, ma'am?" After several seconds, the rescuer reported, "She's not responding, sir."

A closer comview revealed the figure was Astraea. Fidelis clenched his teeth. His face tightened with despair as he fought back tears. Weakened by the sight of Astraea's body, his knees buckled. He would have fallen if Juvo hadn't held him tightly. Fidelis couldn't believe what he was seeing. Only hours earlier, he and Astraea were in the Templum to be married. Now, her lifeless image was broadcast across the planet from deep space aboard the bridge of a badly mauled battleship.

Across the deck, a surviving crew member moaned. It was Lieutenant Carpo. She had burns across her body. "Captain!" Miss Carpo called. There was no answer.

Two rescuers from the med team rushed into the Bridge. "Tend to the Captain, now," ordered Commander Clades. They immediately went to Astraea's side. One knelt beside her. The other leaned over her left shoulder and checked her vitals. He shook his head sadly.

"For the gods' sake man, does she have a pulse?"

"Sir, I opened the Captain's hand to check for a heartbeat. She's holding a betrothal pin."

Fidelis closed his eyes. He had given the pin to Astraea for their engagement. By custom, she would return it at their wedding. His throat tightened. Tears streaked down his cheeks.

One med team member, spoke, "Commander, I'm getting faint vital signs. The Captain is still alive, but she requires urgent care."

Juvo choked back tears, "Praise the gods! I knew she was alive. I knew it!"

A voice called from the rescue party. "Get the Captain immediate medical attention, now!" Control panels near her body erupted into a bright flare. It was a matter of time before other panels would burst into flames. An electromagnetic arc flashed across the bridge. The ship could soon implode.

"Captain? Where's the Captain?" cried Lieutenant Carpo.

"We're tending to her right now, Lieutenant. Please remain still. Don't move. There are broken circuits throughout the Bridge. Help is coming. We're going to get you out of here."

The comcast switched from the rescue scene to an interview. It was Commander Clades. His face was grim. "They were magnificent," he stated. "The *Vindictus* and her crew stopped a vast enemy armada. They fought over three hundred enemy warships without hesitation. The *Vindictus* and her crew destroyed them all."

A journalist asked what would become of the ship?

"The *Vindictus* will return home with sacred cargo. She carries many heroes who gave their lives in battle.

We're transferring survivors to other ships for immediate medical attention. The *Repulse*, the *Courageous* and the *Furious* have joined us."

"Lieutenant Commander Clades," called another journalist, "I thought Junius Brutus Timens captained the *Vindictus*. Who is Captain Astraea?"

"Astraea? She will forever be a legend," replied Clades. "The Captain led her crew through an impossible battle. Now, she fights for her last breath. We take her home to her loved ones. May our gods be with her."

Juvo turned off his telecom. "She's alive Fidey! Astraea's alive! They'll have her planet-side in hours. You'll get to see her!"

Fidelis was unsteady on his feet. Juvo helped his friend step forward. He walked Fidelis from the Templum. They would wait for Astraea at the Landing Port. Hundreds of others would await their loved ones' arrival there. Sadly, most would instead claim the remains of their sons, daughters, husbands, or wives. War provides no joy, only despair.

[End of Chapter]

Attack of the CHAPTER THIRTEEN

Joker Squadron

As Astraea led the *Vindictus* into battle, others joined the fight. They included Commander Scitus. As news of the Visitors' attack spread, he hurried to his ship, the *GSS Tigress*. However, a Launch Control accident prevented him from boarding her. The Commander's valiant crew and ship fought the 'Visitors' without him. The *Tigress* was lost in battle with all hands. Scitus regretted not commanding her in battle.

Instead, the Commander organized a squadron of six lightly armed galactic cutters to fight the enemy. They were built for speed and reconnaissance. Their only weapons were automatic repeating pulse cannons. The cutters' shielding was too light for a fleet battle. However, the Commander believed they could make a difference in desperate times.

Once his Marine escort left, Scitus shook his head to shrug away memories of the launch accident. Among other things, he saw an airman engulfed in flames. The suffering soul flailed about in pain, trying to fight the fire that fed upon his flesh. It was one of many horrors the Commander wished he had never witnessed. Military life feeds nightmares.

Ahead of him were five young pilots. A tall, rugged looking young man tended to his Cutter. Scitus recognized him. He was Ensign Trepidus. The Ensign's actions during the Barbary Wars were the subject of a Naval Board of Inquiry. The Commander was one of its three presiding officers.

The remaining pilots included three females and one male. They anxiously watched and listened to combat reports on their telecoms. An impressive young woman appeared to be in command. The Commander learned later that she was Ensign Precator. The others followed her directions and sought her advice. The Commander knew she wasn't their squadron leader, although most of her squadmates followed her,

Scitus walked past the gathering. He yelled at them, "Why are you all standing there? Are you waiting for a pizza delivery? Don't you know there's a war going on?"

Trepidus stepped forward. He was surprised to see Commander Scitus. The Ensign remembered him as one of three officers who judged him. The Commander didn't seem to recognize him. That comforted the young officer. He answered, "We're awaiting Orders, sir."

"Launch Control is down, Ensign. There are no orders coming."

Ensign Precator stepped forward. She asked, "Then, why are you here, Commander?"

"I'm here to kill the enemy, Ensign. Why the hell are you here?"

All five Ensigns stood at attention and loudly replied, "We're here to kill the enemy, sir!"

Scitus took command. "Who is your squadron leader?"

Trepidus answered. "I am, sir."

"Ensign Trepidus, you have a pilot missing. Where's your sixth pilot?"

"Ensign Mali is our sixth pilot, sir. She's in route from Launch Control. We expect her any minute." Trepidus searched the Commander's face. He looked for a trace of recognition. He wondered how Commander Scitus could have forgotten a six-hour Naval Inquiry about his unauthorized attack against the infamous pirate ship, *Latro*.

Scitus wasn't concerned about Trepidus. He was a skilled pilot. The Commander didn't expect Ensign Mali's return. Anyone at the Launch Center was likely dead, dying, or gravely injured. The Commander wouldn't tell her squadron mates. They needed to focus on tasks ahead.

"How old are you, Ensign Trepidus?" asked Scitus. The Commander knew the answer. He asked anyway. All the pilots were his own son's age.

"I'm twenty-one, Commander."

"Are your ships armed, Mr. Trepidus?"

"Aye, sir. All galactic cutters are armed and ready for war, Commander."

No one is ever ready for war. Scitus knew that. However, he also knew Trepidus was battle tested. He was no stranger to combat.

"Are your pilots ready for action, Ensign?"

Trepidus confidently replied, "Aye, sir, Joker Squadron is ready for duty."

"Joker Squadron?" asked Scitus. It was an unusual name.

"Aye, sir. Joker Squadron is always on deck, ready for action."

"Are you trying to be funny, Ensign?"

"No, sir. That's the squadron motto."

"Well, Mr. Trepidus, I like it. Joker Squadron mount up. We've got a war to fight. Ensign Trepidus, you're my Second in Command."

One of the Ensigns stepped forwards and spoke. "Commander, I'm Ensign Scurra. You might not want Trepidus as your second."

Trepidus paused. Scitus wondered if they knew about the *Latro* incident. "Why do you say that, Miss Scurra?"

"I went to the Academy with Trepidus, sir. He's a troublemaker."

Neither Scitus nor Trepidus wanted to hear about the *Latro* Inquiry. It was a closed matter that wasn't open to public scrutiny. Its records were sealed. Curiously, the Commander inquired, "Ensign why do you say that?"

"I seen him making trouble, sir. It was before final exams."

Scitus was relieved. Scurra's complaint had nothing to do with the *Latro*. "Yes, Miss Scurra, what about it?"

"Commander, Mr. Trepidus had an exam in the morning. So, we went to a local bar to get drunk. That's how he passed all his tests. It was better than puking. Anyway, before Ensign Trepidus has his first drink, he leans over to this big fracking Marine. He says to him, 'Wanna hear a Marine joke?' The Marine looked at him funny. Then, he warns Mr. Trepidus that he's a fracking Marine! The jarhead stands up real angry and says the two guys next to him is also Marines! They were riled up, too! The first Marine sort of growls at Trepidus and asks, 'You wanna tell that joke to all three of us?"

"How did Trepidus answer, Miss Scurra?"

"Ensign Trepidus says, 'Nah, I don't want to have to explain it three times."

The other Cadets chuckled. They had heard the joke before. So, had the Commander. It was an old favorite, but he enjoyed hearing it again. "Thank you, Ensign Scurra. It's settled then. Mr. Trepidus is my second. Please don't make me explain it four times, understood?"

"Aye, sir."

Trepidus ordered, "Joker Squadron, mount up."

They swiftly boarded their ships. Scitus took Ensign Mali's Cutter.

"I need a Navigator. Does anyone volunteer?"

Ensign Avens buzzed in, "Commander, I do navigation, sir."

Scurra spoke again. "Commander, you might not want Avens as your navigator, sir."

"You're not serious are you, Ensign?"

"Sir, you need to know. Ensign Avens was in my class at the Academy. I seen him navigating. We was flying, sir. The pilot was scared when he saw Avens was his navigator. He knew Avens was crazy and would get us lost."

Scitus knew where this conversation was going. He listened anyway.

"So, the pilot pulls out a pistol and lays it on the instrument panel. He growls, 'Mr. Avens, if you get us lost, I'll shoot your ass!' Mr. Avens just smiled. Then, he lays this BIG pistol in his lap. The pilot says what's that for? Avens says, 'With due respect, sir. If we get lost, I'll know before you."

The other pilots fought back laughter, including Avens. It was another old joke, probably the most popular navigator tale in the Navy. The Commander was glad to hear it. He understood how humor eased anticipation of war.

“Avens,” replied Commander Scitus, “you’re our navigator. If you get lost, please don’t shoot anyone.”

“Aye, sir.”

“Ensign Avens. Set a course for Sector 16. I understand there’s a War there.”

“Aye, sir. I’m setting coordinates now.”

Scitus continued “I need a communications officer. Is there a volunteer?”

A voice cleared the comm. “Commander, this is Ensign Statera. I volunteer, sir.”

There was a pause. Everyone expected Scurra to butt in, and she did.

“Commander, this is Ensign Scurra. You might not want Statera being your communications officer. I’m serious this time, Commander.”

“Please state your objections, Ensign.” Before Scurra could reply, Statera interrupted. “Commander, I’m receiving an urgent report from Fleet Command.”

The time for jokes had ended. “Commander, this is Ensign Scurra. Permission to be tail-end Charlie?”

“What are your qualifications to be our tail-end Charlie, Miss Scurra?”

“Class rank, sir.”

Scitus admired the Ensign’s sense of humor. For a moment, he wondered whether Joker Squadron was named for her.

“Ensign Scurra, you’re aware the last vessel in formation is most at risk?” “Aye sir.”

Scurra obviously had guts. The group was comfortable with her guarding their six. “Permission granted. Joker Squadron, let’s go.”

“Commander, this is Ensign Precator. Permission to serve as your wingman, sir?”

“Can you fly, Ensign?”

“Like a Falcon, sir.”

“Permission granted, Ensign.”

Precator was an outstanding pilot. She graduated near the top of her class at the Academy. Her squadron mates had expected she would be their squadron leader. Instead, Ensign Trepidus showed up one morning with that assignment. No one knew him, nor why he was assigned to Joker Squadron.

Within moments, the cutters blasted into space. They were hours from Sector 16. Fellow servicemen were dying at Outpost Semotus. They hoped to save as many as possible.

Scitus directed Ensign Statera to raise communications with Semotus. He was unsuccessful. There was nothing. He received only dead bandwidth.

“Have you received communications from the fleet, Mr. Statera?”

“None, Commander. I don’t like it, sir. Semotus is our communications hub. We can’t have Fleet communications without it. There’s something wrong.”

“Commander?” asked Avens. “What will we do when we get there?”

Ensign Precator spoke, “Sir, I have classmates at Semotus. I want to fight. I want to kill the enemy, as many as possible.” Her opinions carried much weight amongst her flight mates.

“Roger that,” confirmed Ensign Avens. “We need to give them a beat-down.”

Scurra added, “We all got friends there, Commander. Permission to kill the frackers without mercy.”

The squadron wanted to fight. Only Ensign Trepidus was silent. There was a consensus among the others. They all had war on their minds.

Joker Squadron neared their destination. The Commander issued orders. They would drop out of Faster than Light Speed (FTL). Then, the squadron would quickly evaluate their situation. If there were enemy warships present, they would race away.

Their Cutters were designed for reconnaissance, not battle. Joker squadron’s weapons couldn’t penetrate a warship’s hull. Their shielding couldn’t withstand a single cannon shot. Yet, vessels of war couldn’t match their

Cutters' speed. The coming war offered them a new purpose.

Avens requested permission to engage any enemy vessels at the Outpost. Commander Scitus declined. That didn't sit well with other pilots. They wanted to fight, not run.

"Commander," announced Avens, "we've reached Sector 16." The squadron dropped from FTL. What they found was disturbing. There were no enemy warships. Outpost Semotus was smoldering in ruins. Communications from the base were absent. There were no pleas for help. It was deathly quiet.

Ensign Statera reported, "There's nothing down there, sir. No one is alive."

Avens broke in. "Commander Scitus, I'm picking up vital signs. They're feint, sir. There may be a survivor down there. I've got coordinates. Permission to land and evacuate, sir?"

"Permission denied, Ensign Avens. We need you here. I will investigate. There's a broken launchway remaining at the Outpost. Ensign Trepidus and I will land there. I want the rest of the squadron to remain in orbit. Gather information. Keep me appraised."

"Aye, sir," Avens replied.

Scitus and Trepidus broke away from the squadron. The remaining pilots were unhappy with their roles. However, they completed their duties.

Something troubling was discovered.

"Commander, this is Ensign Precator. There was a battle up here. I've located debris from the *Essex.* No survivors are detected, sir."

The *Essex* was one of the Fleet's two Battle Cruisers. Other than the *Vindictus*, she was the Navy's most powerful warship.

"Commander, there's debris from the *Renown*. She was slashed apart." The *Renown* was the fleet's other Battle Cruiser. The only remaining capital ship in the fleet was the *Vindictus*. "There are no survivors, sir. We have debris from merchant ships, and four other Geminum warships. I'm sorry, sir, but one is the *Tigress*. The others include the *Renown*, *Invincible* and the *Indefatigable*. There are no survivors."

"Commander," Miss Precator interjected, "we had friends and classmates aboard those ships. Someone's got to pay for what those frackers did here. We want payback."

Scitus understood, but it would have to wait. He and Trepidus searched for survivors. The Outpost had limited defenses, which weren't intended to fight warships. Devastation surrounded them. Mangled bodies were strewn across the compound. Every structure, building and hangar had been flattened and burned. As the two men approached a battered gun emplacement, they found four men's bodies. All died at their post. One still held ammunition in his dead hands.

"Commander, I've found someone over here." There was a gravely wounded man in the rubble. He struggled to regain his feet, but he had no legs. It was Chief Nux. "They're coming back," he said. "Help me up. We got to get ready."

Scitus knelt beside the suffering man. "At ease, Chief. Tell me what happened here."

Beleaguered from shock and blood loss, Nux replied. "They just come from nowhere. First, they attacked our merchantmen. There was over sixty enemy ships. Once they finished with the civilian transports, the things attacked Semotus. None of our weapons could hurt them. We hit them again and again. It didn't faze none of them. Nothing did." The Chief grimaced. He handled pain well, but it was fast overtaking him.

Trepidus opened a med pack. He administered heavy doses of pain medication to relieve Nux's suffering. "That'll help, Chief."

Nux grabbed Scitus by the upper arm. "Did they get away? I told them to get the hell away." Scitus asked who, but the Chief couldn't respond. His life was nearing an end. "Gemma is my girl," he told the men. "I never had a family, you know? She's the daughter I never had. I'm proud of her. Did she get away?"

The Commander didn't know who Nux was talking about. However, he understood Gemma was important to the dying Chief. Scitus wouldn't let a fellow serviceman die with regrets. He consoled him. "Chief, Gemma made it out. She's safe."

That eased Nux's anguish. "I knew she was okay. She has my wrench, you know?" With a pained grin, Chief Nux said his last words. "So long as she has it, she'll be okay. You'll see." He continued to grip Scitus' arm, but the Chief was dead.

"Who was Gemma?" asked Trepidus.

"I don't know, but nobody survived here. They all died. The Chief didn't need to know that."

"Aye, sir."

Ensign Statera buzzed the Commander. "Sir?"

"Yes, Ensign."

"We're getting readings out here. I think enemy ships are returning. Something is dropping from FTL. They aren't ours. There are twelve warships, Commander. They're badly damaged."

Chief Nux was right. The enemy were returning. However, they weren't looking for another fight. A dozen enemy vessels were badly damaged from their fight with the *Vindictus*. The limping wrecks were the only survivors from the first attack against the dreadnaught. Each was pathetic. All were gashed, shredded, and scorched.

Scitus and Trepidus rejoined the Squadron. "Mr. Avens," ordered the Commander, "place us on an intercept course with the surviving enemy warships."

"Aye, sir. We will obliterate the enemy."

"They'll regret their actions," added Miss Precator.

Scitus ordered otherwise. "No, we won't engage them. Keep a discreet, safe distance. We'll follow them to their base. Let's learn who our enemy is."

"Aye, sir. Intercept course plotted." The enemy warships were easily detected. They leaked plasma, ions, and debris. Most had erratic heat signatures from compromised propulsion systems. "Commander," called Avens, "most of the enemy ships are emitting lethal levels of radiation.'

"Sir, one of them is falling behind. She's heavily damaged. Her systems are shutting down. Life support is failing. The other ships are leaving her behind. She's adrift without power. Commander, that ship has broken apart."

Its debris erupted into a brief, violent fusion blast.

"There are no survivors. No debris. She's gone, sir."

Moments passed before another enemy ship fell from formation. She drifted erratically with an uneven pace. The scarred remains of the vessel's head resembled a ghastly, grinning skull. It was trailed by the ship's smoldering, broken skeleton. Plasma leaked from her gaping wounds into the expanse. She drifted further behind her squadron as her propulsion systems sputtered. The vessel's lights flickered erratically before suffering a massive power surge. When her innards darkened, the trembling eidolon burst apart in a blinding blaze. The remaining ten derelicts shambled forward in a forced funerary procession.

"Commander," signaled Precator. "I think I speak for the squadron when I tell you that we want to attack and kill the enemy."

Avens requested, "Permission to attack the enemy?"

Ensign Statera added. "We want justice, sir."

The Commander understood their emotions. Instead of repeating his order, Scitus queried, “What say you, Ensign Trepidus? You’re Joker Squadron’s Leader.”

Trepidus was silent for a moment. Then, he replied, “Ensign Precator is right, Commander. Retribution is required, but not now. We need to follow those ships to their base. That’s the only way we’ll know who sent them.”

Commander Scitus agreed.

Scurra observed the enemy stragglers. She commented, “Look at them. They came upon us as fierce specters. From nowhere they appeared and viciously attacked innocents. Now, they drag themselves through their own blood. They pretend they aren’t dead. Yet, their final odyssey takes them home as “Phasmii.” They will forever be branded by shame and defeat.”

The Commander replied. “Miss Scurra, you’ve named them well. Hence forward, we shall call them ‘Phasmii.”

Each member of the Squadron voiced approval. “Aye, sir.” The name was later universally accepted by the military. Prime Minister Morus also preferred to call them "Phasmii," instead of his "Visitors."

An hour passed. There was another unexpected development. Statera alerted Commander Scitus. "Commander, sensors read another eight vessels ahead."

"Can you identify them?"

"No, sir. They're different from the Phasmii. They carry unusually high levels of baryon particles. I'm getting increased power readings from them. They appear to be warships, likely destroyers. Sir, each squadron is maneuvering towards the other. They're powering weapons."

Within moments, the mystery ships unleashed fierce cannonades against the Phasmii. Their targets erupted into brief, bright, brilliant blazes, which quickly darkened to nothingness. The strange ships powered down their weapons and withdrew.

"Sir, the Phasmii are gone. There's nothing left of them. No debris. No bodies. It's like they never existed."

Mr. Avens added, “The others have changed course. They’re moving away. Should we follow?”

“No, Ensign. We have one enemy. We need no more. Let’s go home.”

[End of Chapter]

CHAPTER FOURTEEN

Conspiracy of the Humiliated

The Phasmii had been defeated. It was time for jubilation. However, many senior officers were humiliated by the stunning victories of their junior officers. They loathed the success of their world's brave, young heroes. It had cost them financial opportunities, book deals, and cheapened their reputation. To become relevant again, a cabal of high-ranking officials intended to minimize the heroism of those who fought the Phasmii menace. They planned to use a Naval Inquiry to trivialize their deeds and memories. Captain Astraea was their main target. They would make the young heroine appear to be a bungling, incompetent junior officer who nearly snatched defeat from the jaws of victory. They would blame Astraea for the loss of the fleet's mighty dreadnaught.

Many military families understood what was happening. An angry crowd stirred in front of the Campos Naval Courthouse. Shore Police set up barriers to keep protestors a safe distance away. News broadcasts described the Inquiry as a "volatile situation." Protestors gathered. Everyone knew it was a kangaroo court. Powerful, jealous, angry men would try their heroes.

The Inquiry was chaired by Vice Admiral Timens. He was Captain of the *Vindictus* before the Phasmii attack. Timens was absent from the dreadnaught during the Battle. He was severely wounded while boarding a launch which would have taken him to the *Vindictus.* He never boarded his ship. His First Officer died in the same tragedy.

Lieutenant Astraea replaced Timens as Captain of the dreadnaught that saved Gemina from invasion in the greatest battle ever fought. While she and other junior officers earned fame and glory, Captain Timens lay helplessly in a hospital bed. Many believed the senior officer resented Astraea's fame and memory. If not for the Launch tragedy, he would have commanded the *Vindictus* to victory. Timens would be the world's hero. Instead, few remembered his name.

The other Commissioners were Admirals Pavidas and Pullum. They were commonly reviled as cowards. As the two entered Courthouse Square, protestors loudly ridiculed them. Some made chicken clucking calls to the Admirals. Protestors walked with their hands on their sides and elbows behind them. With every step, their heads bobbed forward and backward. "Brrrrawk, brrrrawk," they squawked. Cursing erupted elsewhere. "Cowards! Court-martial the cowards!" An angry woman stepped up to Pullum and spit on him. She said, "My son was aboard the *Essex*! You abandoned him and the rest of your crew when they needed you most. Shame on you, Mr. Pullum!"

Voices deep in the mob yelled, "Is that Pullum? Coward! Coward! Brrrawk, brrrrawk bawk." A few hollered, "*You* should be court martialed, Pullum! Prison for Pullum. Prison for Pullum!"

Others hollered at Pavidas. "Pavidas, ain't got gravitas," they repeated. "Pavidas ain't got gravitas." They booed the two men loudly. As the crowd grew, they jeered in unison, "kangaroo court, kangaroo court, kangaroo court." Military police escorted the two flag officers safely into the courthouse. Many journos referred to the proceedings as a "managed" Inquiry.

Timens approached the crowd alone. He waved away his escort. The Vice Admiral climbed the base of a nearby statue of Justicia. He stood below Lady Justice and shouted. "Hear me, I am Junius Brutus Timens. I was Captain of the *Vindictus* before the attack on Semotus." Clucking began again, until a voice cried out, "Wait, let him speak!" The Vice Admiral continued, "It was my crew who fought aboard the *Vindictus*. I knew each of them. Many lost their lives fighting for us! Others came home crippled, burned, gravely wounded." Timens looked to the mother of the *Essex* fatality. "There were many heroes that day," he assured her. "Your son was among them. I was not. My crew fought without me."

Vice Admiral Timens called out to the gathering. "While your children, spouses, and loved ones fought a vicious, merciless enemy, I lay in a hospital bed. Now, I am here as their voice. There will be justice. That I assure you."

"Astraea," chanted the crowd. Her name was repeated reverently in unison. Others applauded and stomped their feet. Many yelled, "We want justice. We want justice!"

Timens stepped down and walked into the gathering. Many approached him. They told the Vice Admiral the names of their loved ones who died in battle. He knew many of the fallen heroes. He carefully remembered every name mentioned. The Admiral promised all who came to him, “I will never forget. We will never forget.” The crowd escorted Timens to the courthouse steps. Shore police stepped aside.

Pavidas and Pullum met Timens inside the doorway. “What the hell was that Timens?” asked Pullum. However, Pavidas appeared quite satisfied. He gave Timens a wink and smile. He grabbed Pullum and took him aside. Pavidas whispered to his fellow conspirator, “Wise up, Pullum. Timens is just giving them what they want. The more the public loves Timens, the harsher his judgment against the glory hounds.”

Bailiffs ushered the Admirals to their chambers behind the grand courtroom. Each of the men had an agenda. None involved the tactics or strategies of the desperate battles of Sector 12 and 16.

Pullum intended to use the Inquiry to escape mortification. The newly promoted Admiral knew he was an imperfect man. He did many shameful things. He justified them as being necessary to protect his wife and son. Among others, he had framed his son's girlfriend, Cassia Gemma, with a bogus felony prosecution. The falsely accused teen was sentenced to ten years in a maximum-security prison. Unknown to Pullum, his son secretly aided her escape.

Pullum was publicly despised as a coward. Indeed, he was afraid when the Phasmii attacked, but not of death. His fear was that he would widow his wife. He didn't want to leave his son fatherless. The Admiral tried to convince himself that what he did was justified.

Yet, Pullum never did so. He regretted his decision, but there was no way to undo it. It gnawed upon his soul. Shame stared back at him from his mirror. He saw it in the eyes of his wife and son, Amator. Pullum needed redemption desperately. The Naval Inquiry was not where he would find it.

Pavidas was different. The sour old man had no remorse. He was an unrepentant, arrogant admiral. He was furious at the young men and women who dared fight off the Phasmii invasion. They made him look small. To

Pavidas, the Naval Inquiry was an opportunity to regain his once lofty stature. Once again, he would be a giant. Through the Inquiry, he would brand Astraea and other martyred defenders as irrelevant, bumbling dolts.

Pavidas believed Admiral Timens shared his agenda. He believed the former *Vindictus* Captain was a man jilted by fate. If not for the terrible launch accident, Timens would have been the hero of The Battle of Sectors 12, instead of Lieutenant Astraea. She stole his fame and glory and made him irrelevant. Pavidas had no doubt that Timens would be an invaluable ally. He would help humiliate Astraea and other "glory hounds.

Admiral Pullum viewed Timens much the same way. He believed Astraea humiliated her former Captain. In his place, she commanded his ship to victory in history's most epic battle. Pundits often wondered what would have happened if Timens had commanded the powerful dreadnaught in the Battle of Sector 12? Many argued he wouldn't have defeated the Phasmii hordes. Some critics unfairly branded him, "Timid Timens."

As Pullum and Pavidas chatted, Vice Admiral Timens washed his hands. Beyond his hearing, his fellow

Admirals discussed Timens' role in the Inquiry. Pavidas sneered, "Those loud idiots outside think Lieutenant Astraea was a great leader. They will learn differently. There was no chance of her losing that battle. She bungled her way to glory."

Pullum was surprised. "Why do you say that?"

"During the battle, I spoke with Timens when he was in the hospital. Everyone at headquarters was hysterical about what happened at Outpost Semotus. They wanted to know if we had a chance of fighting off the Phasmii attack. I got a nurse to awaken Timens for me. He was heavily medicated, so he was groggy. I let him know everything was going to hades. He said not to worry. Timens nodded to me and said the *Vindictus* had a 'secret weapon.' He told me with certainty that the attackers were doomed."

"There was a secret weapon? Did Timens tell you what it was?"

Pavidas replied, "No, he just said not to worry. Then, he nodded off. The miserable rascal fell asleep from pain killers."

"I see."

Pavidas concluded, "Had Timens commanded the *Vindictus,* the dreadnaught would have returned as a proud champion. She wouldn't have been towed back as a broken funeral scow.'

"By the end of our Inquiry," continued Pavidas, "the world will rue the day that Lieutenant Astraea took command of Timens' ship. Finally, our humiliation will end. Astraea and her glory hounds will be ridiculed and condemned instead of us." Pavidas laughed gleefully. He would regain his fame and fortune.

The Inquiry began an hour later. Timens was surprised by the witness list. He looked at his fellow Admirals. "Gentlemen, I thought this was to be an Inquiry into our Naval tactics and strategies in the Battles of Sectors 12 and 16?"

Pavidas explained, "No, Timens, this is our opportunity to set the record straight. Excuse me. I meant to say, 'Vice Admiral Timens."

"Yes," Pullum added, "the Prime Minister wants citizens to understand the perils of interplanetary war."

"Gentlemen, are you saying that Prime Minister Morus would have us humiliate those who fought and died in our battles against the Phasmii attack?"

"We know what Morus wants, Timens. We understand how Astraea stole your glory. You should have been the hero of Sector 12, not her. All of us have been humiliated and reviled since then. All that will end with this Inquiry."

"Is that why I was promoted to Vice Admiral?"

"Of course. You were the real captain of the *Vindictus*. As Chairman of this Court, you give it authority we wouldn't otherwise have."

"I see." Timens replied.

The three men donned their black robes. They entered the Courtroom. It was a large wood-paneled chamber with sixty rows of pews. Hundreds attended. Many journos raced back and forth for advantageous positions. They knew the world hungered for news about the heroes who defeated the Phasmii.

Civilians crowded into the room, as well as stellarii and other service members. A raucous atmosphere filled the place. Admiral Timens sat between the other Admirals on a judicial bench above the onlookers. Pavidas and Pullum eagerly awaited the proceeding by his side.

An elegantly dressed Sergeant of Arms stepped forward. He saluted the Admirals. As they returned his salutation, the Sergeant announced, "All rise. Oyez, oyez. This Naval Board of Inquiry of the Battles of Sectors 12 and 16 shall hereby convene. The Honorable Admirals Timens, Pavidas, and Pullum presiding."

Everyone took their seats, except for journos. They scurried from place to place, seeking interviews. One of them broadcast a headline for the events as, "Inquiry or Kangaroo Court? The Trial of Captain Astraea Begins."

[End of Chapter]

CHAPTER FIFTEEN

The Naval Inquiry Begins

Gentlemen," asked Vice Admiral Timens, "who will you call as your first witness?"

Pavidas answered, "I thought we should call Lieutenant Commander Negatio."

"But Negatio was one of Astraea's officers," complained Pullum. "She promoted him to Lieutenant Commander. She trusted him. How will his testimony help us?"

Pavidas, answered. "Negatio despised Astraea. He knew she was unfit for command of the *Vindictus*. The Commander didn't want a giddy bride as Captain of our only dreadnaught. Negatio wanted a fighting man in control. He demanded that the weapons officer lead the *Vindictus* into battle. I believe that man's name was

Telum. It was Negatio that tried to lead a mutiny against Lieutenant Astraea. His testimony will destroy her."

"I see." Pullum replied. He turned to Timens and asked, "You were Negatio's Captain. As you told the crowd outside, you knew every member of your crew. Tell us what Negatio will say."

The former *Vindictus* Captain answered, "You won't know until you ask him."

Pavidas laughed. His first witness would paint Astraea as unfit for command. "The Inquiry calls Lieutenant Commander Emendare Crispin Negatio," announced the irate Admiral.

The Sergeant of Arms announced, "This Inquiry calls upon Lieutenant Commander Emendare Crispin Negatio to testify. Please come forward, sir."

Negatio arose from his seat. He straightened his uniform and approached the witness box. The court reporter turned to him. "Mister Negatio, do you swear to the gods, or any god, that the testimony you are about to give is true to the best of your knowledge and belief?"

"I do."

"Please be seated. State your name for the record, sir."

Negatio took a deep breath. He didn't want to remember events from Sector 12. They were traumatic, overwhelming. However, he would not shirk his duty. The Commander answered. "My name is Emendare Crispin Negatio. I'm a Lieutenant Commander in Gemina's Galactic Navy."

Admiral Pullum began the questioning. "Sir, you were the Damage Control Officer of the *Vindictus* during the Incident at Sector 12?"

Negatio found it odd to call the desperate Battle of Sector 12 an "incident." However, he answered, "I was."

"When did you first learn *Lieutenant* Astraea would command the ship?"

"I learned she would captain the *Vindictus* shortly after boarding, sir. We had a meeting of department heads."

Pavidas relished his opportunity to portray Astraea as a shallow, young woman who had no business commanding the *Vindictus*.

"Mr. Negatio, isn't it true that you objected to Ms. Astraea having command of the *Vindictus*?"

"Yes, sir. It was my opinion that the ship's weapon control officer should lead us into battle."

"Did you express that opinion, Lieutenant Commander?"

Negatio was uneasy about the Admiral's questions. "Sir, I was an inexperienced department head. War overwhelmed us without warning. My opinion had little value."

"Answer my question, sir. Did you object to Astraea having command of the *GSS Vindictus*?"

The Commander didn't answer. For a moment, he suffered a flashback. Negatio remembered Astraea's surprise when Lieutenant Carpo announced, "Captain is on the Bridge." Without hesitation, Astraea adapted to the situation and took command. She focused immediately. There was nothing frivolous about her.

Pavidas raised an eyebrow and stared down Negatio. "Pay attention, sir. Let me remind you this is a Naval Board of Inquiry. Isn't it true that you spoke against Lieutenant Astraea having the ship's command?"

Mr. Negatio buried his face in his hands. He slowly raised his head and answered, "Admiral, I spoke in an ill-tempered manner. I was nervous. We all were. The *Vindictus* was headed into war without our Captain or First Officer."

Negatio looked toward his former Captain, Admiral Timens. "Sir, we weren't ready for battle without you."

"You proved otherwise, Mr. Negatio," Timens replied. "I wish I had been aboard. I apologize for my absence."

Admiral Pullum interrupted. "Commander Negatio, answer the question. Did you speak against Lieutenant Astraea commanding the ship?"

After a long pause, Negatio answered, "I did, Admiral."

"Didn't you complain about the Lieutenant's attire?"

"I did, sir."

"Didn't you say that Astraea fled to the *Vindictus* to escape her wedding?

Mr. Negatio sat quietly without answering.

Pullum continued, "Wasn't her groom some local politician named *Fidelis*? I take it you didn't vote for the man?"

"I spoke improvidently. I wish I could withdraw my remarks. I will always regret them."

Admiral Pavidas enjoyed intimidating the witness. He wrinkled his face and snarled, "Commander, isn't it true that Lieutenant Astraea strolled aboard the *Vindictus* wearing high heels and white lace? Was she pretty, Commander?"

"Admiral, she was called to duty from her wedding. What I said was improvident, inexcusable and wrong."

"You didn't answer my question, Commander." Pavidas scowled, "Was Lieutenant Astraea pretty when she took command of the fleet's only dreadnaught?"

"Are you questioning the Captain's attire?" asked Negatio.

Pavidas demanded an answer. "What was the Lieutenant wearing?"

"She was dressed for her wedding, sir. The Captain wore her dress uniform, a golden Cuirassier helmet with laced plume, three-inch heeled boots, and laced armorettes."

"Was that appropriate attire to command a dreadnaught?"

Mr. Negatio was surprised by the question. He replied, "She didn't dress for combat, Admiral." After a brief pause, he asked the Admiral, "Did Mrs. Pavidas wear combat gear to *your* wedding, sir?"

There were chuckles from the gallery. Someone whispered loudly, "She should have."

Admiral Pullum rigorously tapped his gavel. It was his turn to question Negatio. "Commander, you not only criticized the Lieutenant's party clothes, but you also challenged her tactics. Isn't that true?"

"It is, sir. I disagreed with the Captain's decision to take the ship into Sector 12. It was a 'No-Go Zone.' I also disagreed with her decision to anchor the *Vindictus* in battle. I believed her commands were imprudent."

"Of course, they were, Commander," replied Pullum dismissively. "Every plebe at the Academy is taught that battles are won by outmaneuvering the enemy. Astraea's foolish order denied that advantage to the *Vindictus* and her crew. It was a fundamental error."

"Admiral," continued Mr. Negation, "I didn't understand her orders at the time."

"Did Lieutenant Astraea explain herself?"

"No, sir. Captain Astraea said her duty was to command, not to explain her orders."

"So, this young woman came aboard the fleet's most powerful warship dressed for a wedding party? She took command after her friend, Ensign Carpo, announced her as the ship's 'Captain?' Then, she ran the ship incompetently?"

Negatio, reluctantly answered, "Initially, I questioned Captain Astraea's orders. I didn't know about the quantum tide."

Pavidas remarked, "There is no such thing as a 'quantum tide,' Mr. Negatio. You can speak freely, Commander. Tell us about the ship's 'secret weapon."

"Secret weapon, Admiral?"

"Yes, Commander. Tell us about the secret weapon. Why didn't Astraea use it when the enemy first came upon you? Why did she wait until the *Vindictus* had suffered grievous damages?"

"I don't know anything about a 'secret weapon,' Admiral."

Pavidas turned to Admiral Timens. He whispered, "It's your turn. Finish him off. Negatio was one of your crewmen. He's used to following your orders."

Timens leaned forward. "Mr. Negatio, thank you for being here. I remember when you were first assigned to the *Vindictus* as an Ensign. You were fresh from the Academy."

"Aye, sir. It was an honor to serve with you."

"Admirals Pullum and Pavidas have raised several questions about Miss Astraea. Would you mind answering a few questions for me?"

"I will gladly answer, sir."

Pavidas sneered. He winked at Pullum and elbowed him slightly in the ribs. "Now," he snarled, "Timens will destroy the boy. Watch."

Timens continued, "You did an outstanding job at Sector 12, Commander." The Admiral turned to the court reporter. "Please note for the record that Lieutenant Commander Negatio demonstrated great courage and leadership in the battle against the Phasmii. Although his ship suffered crippling damages from enemy fire, Mr. Negatio kept the *GSS Vindictus* intact until the enemy were vanquished."

The reporter carefully recorded each word.

Mr. Negatio replied, "Captain Timens... Admiral, I didn't do it myself. Each member of the crew kept the *Vindictus* fighting. Most lost their lives. Most survivors were severely wounded. All of us did what we could. It wasn't *me*."

"Of course," Timens replied. Then, he added. "Madam Reporter, please note my nomination of Mr. Negatio and every crew member of the *Vindictus* to receive the Navy Medal of Valor."

Pullum and Pavidas appreciated how Timens had disarmed Negatio. They believed the Vice Admiral had readied Negatio to humiliate Astraea.

Timens proceeded. "Commander, you were asked about the 'secret weapon."

Negatio was puzzled. He answered, "Yes, sir?"

"Let's not discuss that now. I understand that Lieutenant Commander Clades will testify later. He shall address it. Let us instead talk about what happened at Sector 12. Was there ever a time when you believed the ship was lost?"

"Many times, sir."

Pavidas nodded at Pullum. He was satisfied with Timens' mastery of a former subordinate.

"What kept the *Vindictus* fighting, Commander?"

"Captain Astraea, sir. The crew had confidence in her. She was clever, confident, and brave. We trusted her. When her tactics beat back the first wave of Phasmii attackers, no one doubted her. She was brilliant. When the second wave came, there was little hope of us surviving, much less driving them back. There were over two hundred more Phasmii ships. We were doomed. However, the Captain told us to have faith in her. She promised we would kill them all. We knew her word was true. Every one of us fought beyond the point of exhaustion and pain thresholds. Even the wounded and dying fought. The ship was wrecked, but we fought. The Captain said 'give me 14 more minutes, and they will all be dead. Gemina will be saved.' We gave her everything we had. For us survivors, it was fourteen minutes. For others, it was their life."

"Then what happened?"

"The quantum tide came from the asteroid belt. It rushed forward at great speed and swamped the Phasmii. The tide pulled them into the asteroid belt. Then, it spit the frackers out and sucked them back in. It did it dozens of times, until there was nothing left of them, except space dust. As Captain Astraea promised, they were all dead. Their ships were ground into space dust. Gemina was saved."

Pavidas and Pullum were upset by the Commander's answer. It wasn't what they expected. They loudly tapped their gavels. "There's no such thing as a 'quantum tide, *none*," snarled Pavidas! "It's a cover-up for the secret weapon!"

Admiral Pullum intervened. He stared at the witness and scolded Mr. Negatio. "So, Lieutenant Astraea simply snapped her fingers in her pretty outfit, and this 'quantum tide' appeared from nowhere and swept away hundreds of enemy warships? Is that your testimony?"

"That's not what I said, Admiral."

Channel One quickly broadcast from the Inquiry, "Pavidas claims no such thing as 'quantum tide.' Reveals a secret weapon was used against Phasmii at Sector 12. Developing."

As Chairman of the Inquiry, Admiral Timens resumed his questioning.

"Mr. Negatio, were you satisfied with Captain Astraea's command?"

"I was, sir. If not for Captain Astraea, the Phasmii would have overrun our defenses. They would have destroyed Gemina. She was a hero, sir. You would have been proud of her."

Timens nodded his head. "You served Captain Astraea well. If she were here, I know she would thank you, sir."

The Commander began to rise. However, Admiral Timens asked him to remain seated. "I have one *last* question, Mr. Negatio."

Pavidas and Pullum leaned forward. They believed Timens had saved the most savage question for last.

Timens continued, "There was mention that Lieutenant Astraea boarded the *Vindictus* wearing high heels and white lace. Have you anything further to say about that?"

"Yes, sir. She would have made a beautiful bride."

Timens responded by looking at his fellow Admirals. He replied with a wry smile, "Commander Negatio, some of us wear lace better than others."

Several chortles broke out in the gallery as spectators imagined Admirals Pavidas and Pullum in lace.

Negatio grinned, "No offense, Admiral, but none of you could have pulled it off like Captain Astraea."

Again, there were scattered chuckles in the courtroom. Admiral Timens approvingly smiled and dismissed the Commander. "Thank you for your service, sir. May the gods protect you always."

Pavidas angrily tapped his gavel and called a recess. He and Pullum gathered privately with Timens in Chambers.

Pavidas was furious. "What's wrong with you Timens? You had a chance to humiliate Astraea. You know that run-away bride should have used the secret weapon before the Phasmii savaged your ship? She bungled the whole thing."

"So, you want me to publicly disclose our 'secret weapon?" queried Timens.

"Of course, that's the only chance to reclaim our lives. Those 'heroes' humiliated us. I've lost book deals. The networks don't want me. You, me, and Pullum will remain irrelevant until the public learns how their 'heroes' nearly blundered away victory."

Admiral Pullum sourly added, "Timens, get serious. This is our chance to win back public respect. You need to focus."

Timens queried, "You seek respect by humiliating our young heroes?"

"We were the heroes," Pavidas complained. "Our next witness is Lieutenant Carpo. She's another one of your crewmen. Her testimony will crush Lieutenant Astraea. Lieutenant Carpo anointed the party girl to be the ship's Captain. Use your authority over her."

Pullum sneered. "This recess is over, Timens. Let's get back to the courtroom. Let's prove that Lieutenant Astraea was an incompetent usurper."

When the three men returned, the gallery rose to their feet. "This Inquiry is in session. Be seated," announced the Sergeant of Arms. Admiral Pavidas called Lieutenant Carpo to testify. She was duly sworn and seated in the witness box beside the Admirals' bench.

"Lieutenant," the sour old man groaned. "You served aboard the *Vindictus*?"

"Aye, sir."

"You were Lieutenant Astraea's Communications Officer during the incident at Sector 12?"

The Lieutenant hesitated to answer the question. She had never heard the Battle of Sector 12 referred to as an "incident."

Reluctantly, she replied, "I was, sir."

The Admiral rubbed his eyes, as if he were weary or bored. Then, he continued his questioning. "Is it true that

Lieutenant Astraea took command of the *Vindictus*?" Pavidas intentionally referred to Astraea as a 'Lieutenant.' He sought to prove she had usurped command of the dreadnaught. He would establish Astraea as a junior officer who was unqualified for command.

"Yes, sir. I announced her as the ship's Captain when she came aboard."

Admiral Pullum feigned surprise. He interrupted, "So, *you* made Astraea the ship's Captain?"

Pavidas loudly added his indignation. "Miss Carpo, *you* made this young, run-away bride the Captain of the Navy's most powerful ship? Didn't you realize we were at war?"

Pullum indignantly demanded. "By what authority did you make Miss Astraea Captain of the *Vindictus*?"

"Sir, as Communications Officer, it was my duty to acknowledge that the ship's Captain came aboard the bridge."

Pullum continued, "No, Lieutenant, by what authority did you announce Miss Astraea as the ship's Captain?"

"She was the senior officer, sir."

Pavidas impatiently tapped his fingers against the table. He groaned. "Why did you make her Captain, Miss Carpo? There were other Lieutenants aboard. Why did you choose Lieutenant Astraea to be Captain? Were you and she friends? Did the two of you shop together? I bet you have white high-heeled boots like hers in your cabin. Is that so?"

No one in the gallery found the comments witty or amusing. Miss Carpo was offended. She rose to leave.

Pavidas called her down. "Be seated, Lieutenant. This is a Naval Inquiry."

Carpo looked to her former Captain, now Vice Admiral Timens. He nodded his head affirmatively and requested, "please be seated, Lieutenant." The young officer trusted him. She sat quietly and awaited the next question.

"Why" repeated Pavidas, "did you choose Astraea to be Captain?"

"Admiral, I never claimed to make her Captain."

"Then who did?"

Carpo was confused by the questioning. "Sir, naval regulations made her the ship's Captain. Miss Astraea had the highest ranking of anyone aboard. Her naval ranking was third among all Lieutenants in the fleet. No one else aboard was below seven."

"Miss Carpo, Lieutenant Astraea was an astrophysicist. Why not announce Mr. Telum as Captain? He was the ship's weapons officer. Wasn't he better qualified to lead the *Vindictus* into combat?"

"That was not a decision for me to make, Admiral."

"Isn't it true that no one else wanted Lieutenant Astraea to be the ship's Captain?" complained Pavidas. "Weren't concerns expressed by Department Heads?"

"I wasn't present, sir. I believe Mr. Figo was concerned about the ship entering Sector 12. As you know, it was a 'No-Go Zone.'

"Mr. Figo was the Ship's Engineer?"

"He was acting Engineer, sir."

"Did the Ship's Doctor also express concerns?"

"I believe so, sir. Dr. Medens said the Crew didn't want to hide in Sector 12. They wanted to fight."

"How did Miss Astraea answer their concerns?"

"*Captain* Astraea considered their counsel," Miss Carpo answered. "She then commanded the ship."

Admiral Pavidas turned to the Vice Admiral. It was time for Timens to use Carpo to portray Astraea as a frivolous run-away bride. "Vice Admiral Timens, have you anything to ask Lieutenant Carpo?"

"I do."

Pavidas and Pullum watched Timens win Lieutenant Carpo's trust. The Vice Admiral bowed his head and smiled at his former crewman. "Lieutenant, I'm glad to see you again. Thank you for coming here this morning. We appreciate your testimony."

“Thank you, sir. We prayed for you when we learned of your injuries.”

“And I prayed for each of you. All remain in my prayers.”

Pavidas and Pullum eagerly awaited Timens to bully Miss Carpo. It was a moment they anticipated with delight.

As they watched, Timens concluded, “You’re dismissed, ma’am.” Pullum and Pavidas were stunned. Timens didn’t lay a glove on her. He may as well have blown her a kiss!

The Inquiry broke for morning recess. All three Admirals retired to Chambers. Pavidas was livid. “Timens,” he growled. “Don’t you understand what we’re trying to do? You let that freaking Lieutenant Carpo walk away unscathed.”

“She’s not unscathed,” replied Timens. “Miss Carpo nearly died aboard the *Vindictus*. She has yet to recover from her injuries. What is it you would have her say?”

"Maybe, she should have said the *Vindictus* would have fared better with you as its Captain! Wasn't that the point of having you in this Inquiry? Weren't we going to compare your leadership to Astraea's bungling? I thought we were going to contrast an inexperienced Lieutenant with her veteran Captain?"

"Is that what you believe?" asked Timens. No one answered.

[End of Chapter]

CHAPTER SIXTEEN

After Morning Recess

The three Admirals retired to Chambers. Pavidas and Pullum fumed at Timens. "Who is next?" asked Pullum.

Pavidas sneered. "Let's have the AWOL Petty Officer testify. Mr. Iratus will tell everyone how Lieutenant Astraea let him hitch a ride aboard the *Vindictus*." Maybe, she found him more appealing than her jilted groom?"

The misguided Inquiry resumed after morning recess. Admiral Pavidas called Petty Officer Iratus to testify. He was a stranger to Astraea when she allowed him to accompany her aboard the *Vindictus*.

After being duly sworn, Petty Officer Iratus took the witness stand. He was a broad shouldered, well-muscled noncommissioned officer. Iratus sat in the witness box with a defiant frown. Admiral Pavidas first questioned the petty officer.

"Petty Officer Iratus, I understand you were aboard the *Vindictus* during the Incident at Sector 12?"

As had other witnesses, Iratus was confused by reference to the Battle of Sector 12 as an "incident." He reluctantly answered, "I was, sir."

"But you weren't a member of the ship's crew?"

"No, sir. Captain Astraea allowed me aboard."

"I'm puzzled, Mr. Iratus," muttered Pavidas. "You weren't a member of the crew, but 'Lieutenant' Astraea let you board the *Vindictus*?" Pavidas turned to his fellow jurists and cackled, "Perhaps the 'Lieutenant's' parents forgot to tell her not to pick up hitchhikers?"

His witty remark was unappreciated. No one in the gallery laughed. Several groaned. Iratus was irritated.

Admiral Pavidas resumed questioning. "Petty Officer, please explain why Lieutenant Astraea invited you aboard the *Vindictus*?"

The petty officer was surprised by the question. With a smirk, he answered, "I would like to think it was because of my good looks, Admiral."

There was chuckling and laughter in the gallery. Admiral Pavidas frowned. Then, he angrily tapped his gavel and ordered quiet in the courtroom.

"Did I understand you correctly, Petty Officer?" Pavidas retorted. He cupped his ear and asked, "Did you say Lieutenant Astraea invited you aboard the *Vindictus* because of your 'good looks?"

"No sir. I said, 'I would like to think' she did. I wanted to join the *Vindictus* to fight the enemy. There was a war going on. I had buddies fighting on Semotus. So, I requested permission to join the crew. Captain Astraea needed a gunnery officer. I was qualified, so she accepted me."

Admiral Pavidas scowled. "Is that so, Petty Officer Iratus? How would you know whether the *Vindictus* needed a new gunnery officer?"

That was a good question. The gallery quieted to hear the Petty Officer's answer. Iratus replied, "Admiral, I knew the *Vindictus* needed a gunnery officer, because I seen him running away from the Marines. He was yellin', 'No frackin' way!' That was right after I seen you running away from them, sir. You was faster than him and escaped."

There were howls and snickers from the gallery. The Admiral was embarrassed and irritated. He loudly slammed his gavel. "You are dismissed, Petty Officer Iratus." Then he warned, "You will regret your insolent remarks someday."

Before Iratus left the witness stand, Vice Admiral Timens called to him. "Petty Officer Iratus?"

"Aye, sir," he replied.

"Thank you for joining the *Vindictus* crew. I understand you commanded a graviton cannon battery during the battle?"

"I did, sir."

"I was told you destroyed many enemy vessels."

“Aye, sir – me and my gunnery crew.” Members of the gallery whistled and cheered.

The Vice Admiral smiled gratefully and lightly tapped his gavel. “Petty Officer Iratus, thank you for service.”

“It was my honor, sir.”

Iratus stood and saluted Timens. As he left, members of the gallery poked fun at the petty officer by making blasting sounds. After all, he was a gunner. Knock it off, Iratus laughed. He shook his head at their friendly ribbing. Then he was gone.

[End of Chapter]

CHAPTER SEVENTEEN

Secret Weapon

There was angst among the three Admirals. Pavidas thumped his fingers. He whispered angrily to Timens, "That man hitch-hiked aboard your ship. He was insolent to me. Yet, you made him look like a hero!"

Pullum glowered angrily at Vice Admiral Timens. He reminded him that the next witness is Commander Clades. "He's our last chance, Timens. Make him talk about the secret weapon."

Timens nodded affirmatively. "Very well, sir. I will do so." Timens promised to ask Mr. Clades about the ship's secret weapon.

Pavidas added, "Listen, Timens, everyone respects Commander Clades. He's a decorated veteran. He must despise Lieutenant Astraea for running him off the battlefield while the *Vindictus* got all the glory. He would have used the secret weapon before your ship was torn apart by the Phasmii. He will gladly testify about her incompetence."

"Call him, if you wish," replied Vice Admiral Timens.

After being duly sworn, Mr. Clades took the witness stand. He had commanded the *Lioness* in combat against the Phasmii in Sectors 12, 14, 16 and 18. He was a battle-tested Commander, who had organized Gemina's last line of defense. As a reputable warrior, Pavidas expected the veteran Commander would expose Astraea's ineptness.

"Good afternoon, Commander," began Pavidas.

"As you say," grumbled Clades.

"I understand you commanded the *Lioness* during the Incident at Sector 12?"

"The incident, sir? Are you calling what happened at Sector 12 an 'incident?"

"Yes, tell us about the incident."

Clades was unhappy with the Admiral's word choice. "I commanded the *GSS Lioness* at the 'Battles' of Sectors 12, 16 and 18. They were not 'incidents.' It was war. Our planet's fate was in jeopardy."

Surprised by the Commander's answer, Pavidas decided to redirect the witness's testimony. "Didn't Lieutenant Astraea order you and your squadron away from the battlefield?"

"She did. The *Vindictus* remained and fought alone. We weren't allowed to reenter the Sector until 12:05. By then, the Phasmii had been destroyed,"

"So, the *Vindictus* did all the fighting while you watched?"

"She did almost all fighting in Sector 12."

"Why didn't Miss Astraea want your ships in the battle? Was it because you ran away from the enemy in Sector 16, then Sector 14, then Sector 18? Was that why she dismissed you?"

"I don't think you understand," began Clades. "Admiral Forctis led our squadron from Sector 14 into

Sector 12 on a relief mission to Outpost Semotus. The Outpost was under attack by over sixty warships. When we dropped out of FTL, our entire squadron was caught in a swarm of Phasmii warships. They quickly overwhelmed us. Six of our ships were lost before we withdrew."

"Mr. Clades, I understand you were the senior officer among those three warships. Wasn't your first command to run away?"

"We were ordered to withdraw, Admiral."

"You withdrew?" Pavidas sighed with disapproval. "You mean you fled from the enemy again and took two Geminum frigates with you?"

"So, you say, Mr. Pavidas. The *Lioness* fought a rearguard action. Both the *Indomitable* and *Inflexible* suffered substantial damage during the ambush in Sector 16. My ship defended them from Phasmii pursuers." Both the *Indomitable* and *Inflexible* suffered substantial damage during the ambush.

"So, you think you fought off the Phasmii?" queried Pavidas. "No, Commander Clades," he continued, "the enemy had no intention of destroying your ships. They weren't trying to destroy you. The Phasmii wanted you to lead them to the *Vindictus*."

"Admiral Pavidas, the Phasmii attacked us immediately. I led the enemy away from our fleet towards Sector 18. I hoped that would allow our Navy sufficient time to form a defense. However, Captain Astraea ordered us to Sector 12. I followed her order."

"So, Mr. Clades, you took your ragged little squadron to Sector 12 where the remnants of our Navy awaited the entire Phasmii fleet?"

"Captain Astraea ordered us to rendezvous with the *Vindictus* in Sector 12."

The Admiral sighed. "A lowly lieutenant ordered you, sir?"

"Miss Astraea was Captain of the *Vindictus*. She was my superior officer."

"Is that what you were told, Mr. Clades?" asked Pullum.

"It's what I knew, sir."

"Are you telling us that this young woman ordered you to take the *Lioness* and two severely damaged frigates into a hazardous No-Go-Zone?"

"As you say," replied Commander Clades.

"You weren't the only one to flee that day, Commander Clades. Miss Astraea fled her own wedding!"

Clades clenched his teeth. He made a fist and began to rise. The Commander was going to punch the Admiral's face. Never had a witness struck an Investigating Officer at a Naval Inquiry. Yet, there was a first time for everything.

Timens calmly motioned for Mr. Clades to return to his seat. The Commander and Vice Admiral Timens understood one another. They shared a brotherhood of command. Clades resumed the witness box.

"Commander," Pullum continued with contempt, "did you know that Lieutenant Astraea's friend, Ensign Carpo, announced *her* as the ship's Captain?"

In disgust, Clades answered, "I wasn't aboard the *Vindictus*, Admiral." The Commander looked again at

Timens, who quietly motioned for Mr. Clades to suffer Pullum's questions.

"Mr. Clades let me ask you a military question," continued Admiral Pavidas. "Have you ever known a Captain to fight her ship while anchored to a gaseous giant planet?"

Clades understood the Inquiry was misusing his testimony. He didn't like it. "No, sir," he answered. "To my knowledge, no Captain had fought a ship from anchor."

"Wasn't that an obvious tactical error?"

"It was contrary to conventional teaching."

"And that young woman refused your veteran assistance repeatedly?"

"She did, Admiral."

"She didn't want your advice, sir?"

"I cannot speak for the Captain." 'But I can assure you I did everything in my power to protect the fleet.'

Admiral Pullum joined the attack. "Mr. Clades, why did you follow the commands of a run-away bride? She must have been very emotionally distraught!?"

"I followed a superior officer's orders." Clades replied angrily. After pausing, he continued, "Had you been present, Mr. Pullum, I would have followed *your* orders. The crew of your ship, the *Essex,* would have followed your commands had you been aboard her during the ambush at Outpost Semotus. However, you were hiding planet-side with Mr. Pavidas."

Pavidas angrily tapped his gavel. He growled, "Mr. Clades, are you telling us that you were more afraid of Lieutenant Astraea than the Phasmii?"

Mr. Clades had no more patience. "Don't speak to me of fear, Mr. Pavidas. I fought that day in four Sectors. I readied our last defenses. While brave men and women engaged the enemy, *you* hid in a bunker."

The Commander's words drew whistles from the gallery. His reply verged on contempt. Pavidas rose to summon the Master of Arms, but he resumed his seat when his cohort, Admiral Pullum, whispered, "Let it go." Admiral Pullum didn't want to argue his own whereabouts that day.

“Mind you sir,” reminded Admiral Pavidas, “I’m your superior officer.”

Commander Clades retorted, “You commanded no one that day, sir. Do you claim injury? Did you suffer a paper cut, maybe a boo-boo?”

The Commander’s insolence infuriated Pavidas. Someone in the gallery joked, “I think the Admiral got an *ouchie*!” Others laughed. A few made clucking sounds.

Growing exasperated, Pullum lashed out at Mr. Clades. “You are impertinent, sir. If not for our secret weapon, we would have lost the *Vindictus*. The Phasmii would have attacked Gemina. Nothing could have stopped them.”

“Yet, you call it an ‘Incident,’ Admiral?”

“Be careful with your words, sir! Tell us about the secret weapon.”

“What secret weapon, Admiral?”

Pavidas replied. “You know what I’m talking about, Mr. Clades.”

"If there was a 'secret' weapon," replied Clades irritably, "it wouldn't be secret if I told you about it."

Admiral Pullum ordered Commander Clades, "Tell us about the secret weapon, Mr. Clades."

"Admiral Pavidas," answered the Commander, "Don't you know what it was?"

Neither Pavidas nor Pullum knew anything about the secret weapon mentioned by Timens. Pavidas dodged the question. "I'm not at liberty to say, Commander."

Admiral Timens looked at his frustrated colleagues. He smiled at Clades and instructed the Commander, "Mr. Clades, this Commission wants to know what secret weapon won the Battle of Sector 12 and saved our world. Please answer the Admiral's question."

Clades needed a moment to comprehend the request. Once he understood what Admiral Timens intended, he replied, "Very well, Vice Admiral, I will reveal it."

There was excitement in the courtroom. Channel Three broadcast a news flash, "Commander Clades will reveal secret weapon of Sector 12 ... developing."

The courtroom quieted. Pavidas and Pullum eagerly leaned forward. They yearned to hear every word of Clades' answer, as did the gallery and journos.

"Tell us." commanded Pavidas. "What was the secret weapon that obliterated the Phasmii."

"Aye, sir. Our secret weapon was..." Clades hesitated to answer. He asked for the Vice Admiral's reassurance. "May I answer publicly, sir?"

Timens nodded affirmatively.

Clades continued, "Our secret weapon was..." He looked at the gallery. All eyes were upon him. He finished his answer. "The 'secret weapon' was Captain Astraea."

The Admirals were speechless.

Immediately, journos sent a new headline, "Astraea was secret weapon!"

There was loud murmuring in the gallery. Many chuckled.

One yelled, "Surprise!"

Clades continued his answer. "I later learned that when Captain Astraea had been a Science Officer, she was assigned to study Sector 12. During her mission, she discovered the astronomical phenomenon which we now call the 'quantum tide.' That discovery made Sector 12 a 'No-Go Zone' because of hazards it posed to navigation. Captain Astraea lured the enemy into Sector 12 to be destroyed by the quantum tide. She anchored the *Vindictus* to a gravity well to keep the tide's violence from sweeping away her ship with the Phasmii. I was unaware of the phenomena. So were the Phasmii. They learned the hard way."

"Thank you, Mr. Clades," Vice Admiral Timens replied.

Pavidas angrily pounded his gavel. "You are dismissed, Mister Clades. Get out!"

Vice Admiral Timens lightly tapped his gavel. He beckoned to Mr. Clades. "May I shake your hand, sir?"

Clades left the witness box. He walked slowly to Timens. "It would be an honor, sir." The two men exchanged a friendly handshake before the courtroom gallery.

"Commander Clades, thank you for bringing the *Vindictus* crew home. For that, I am forever grateful." Clades' role in the Inquiry had ended. As the Commander left, members of the gallery stood at attention. Many saluted as he passed.

"This Inquiry shall adjourn for one hour," announced Vice Admiral Timens. As members of the gallery exited, there was laughter in the courtroom. Pavidas and Pullum were furious. They scolded Timens in chambers. "You've made a fool of us, Timens. You told me in the hospital there was a secret weapon." Pavidas threw his robe on the floor. "You've ruined us all."

Pullum scolded, "Don't you see, Timens? We're a laughingstock. This Inquiry was our only chance to restore our reputations."

Before Timens answered, Pavidas growled, "You've got your own agenda, Timens. What is it? What game are you playing? Whatever it is, let us in on it."

The Vice Admiral removed his robe and hung it in a closet. He answered, "Mr. Pullum, this is not the way to escape your shame. Admiral Pavidas, how dare you twist the words of young heroes to salvage your ego?"

Indignantly, Pavidas rebuked Timens. "You throw grand words about as if you're relevant, Timens. No one cares about you, Vice Admiral. When you were Captain of the mighty *Vindictus*, you were newsworthy. Don't fool yourself. Despite your rank, you're nothing, now."

Timens sadly shook his head. He left the judge's chambers. Pullum and Pavidas remained behind. "I despise Timens," sneered Pavidas. "He's planning something. I know it. Whatever it is, we're not included. He would abandon *us* to ignominy. Before the incident at Sector 12, I made millions. No longer. The young heroes bask in glory that once paid my bills. Because of them, I have nothing left, except a military pension."

Pullum had a change of heart. He questioned his role in the Inquiry. "Timens is right, you know. If not for those young men and women, Gemina would have been conquered, or worse. My crew perished in battle without me. I see their faces every day. I hear their voices. I miss their company. There isn't a moment when I haven't

CHAPTER EIGHTEEN

Testimony of City Councilman Fidelis

The Inquiry resumed at hour's end. The Courtroom remained packed. Vice Admiral Timens announced the next witness. "Apius Caecus Fidelis, please come forward and testify" There was murmuring from the gallery. It was extremely unusual for civilians to give testimony at a Naval Inquiry. Journos quickly telecast that Councilman Fidelis would testify. No one knew why.

Pullum snickered at the announcement. Admiral Pavidas laughed aloud, "So, let's hear from the groom from whom Astraea escaped by running to war?" They would use Fidelis's testimony to slander the world's greatest hero.

Vice Admiral Timens arose. He was furious. Timens looked down upon the Admirals. "I've heard you both malign Captain Astraea as if you knew her. He slammed his gavel twice, once in each of their directions. "Admirals, you suggested Astraea should never have commanded the

Vindictus – my ship! Hear me now. From my hospital bed, *I* ordered her promotion to Captain."

There were surprised murmurs in the gallery. Journos hurriedly reported what Timens revealed. The two Admirals stared open-mouthed at him.

Timens continued angrily, "You two have behaved disgracefully. I knew Captain Astraea. Now, we shall hear testimony from someone who knew her far better than any of us."

Each of the Admirals began tapping their gavels furiously.

"I'm the Senior Officer of this Tribunal," asserted Vice Admiral Timens. He pointed his gavel at each of the flag officers. "You are both dismissed." He waved them off and commanded, "Leave."

They stared at the Vice Admiral, bewildered. "Go now," he commanded. As they shuffled away in their black robes, Timens firmly struck his gavel against the bench rail and ordered them to hasten their exit. "Move more quickly, or I shall have the Sergeant of Arms assist you."

As Pullum and Pavidas hurried away, the Courtroom doors opened. Unintentionally, Councilmen Fidelis made a gladiatorial entrance as they exited past him. His eyes watered as members of the gallery chanted one name, "Astraea." As all eyes watched, Fidelis walked alone down the courtroom aisle. He was sworn to testify by Vice Admiral Timens.

Mr. Fidelis asked, "Vice Admiral Timens, why have you brought me here?"

"We're here to inquire about the Battle of Sector 12, sir. Some have criticized Captain Astraea for losing the *Vindictus*."

Fidelis was disgusted. "They blame Astraea in her absence? Are her accusers aware the *Vindictus* fought hundreds of enemy warships? Don't they realize the Phasmii were responsible?"

"I wish Captain Astraea could answer for herself, Mr. Fidelis. She was a valued member of my crew. Would you please speak on her behalf?"

"Admiral Timens, Astraea wasn't a Captain to me. She was more than that."

“I understand, sir. Please, tell us what happened that day?”

“I know nothing about the battle, Admiral.”

“Give us your account. That’s all I ask.”

“Very well, it was our wedding day. She stood across the Templum. Astraea was beautiful - the most beautiful woman I’ve ever seen. My friends held me because my knees weakened. That’s when the Marines came. They called her to duty. There were murmurs that war had erupted. She never hesitated. Astraea kissed my cheek and left with them. She promised to return. That evening, the Navy returned her. Astraea was mortally wounded, burned, and irradiated. Her doctors rushed her to intensive care.” Fidelis paused to gather himself.

“She lay comatose. Astraea didn’t know I was by her side. I begged her, “Please come back.” Fidelis’ throat tightened. He swept a gathering tear from his eye before it wetted his cheek.

“I understand your feelings, Councilman.”

"I thanked the gods that Astraea was unconscious. Otherwise, she would have suffered unbearable pain. The doctors said she was dying. They told me Astraea wouldn't survive the night. They suggested I give my final prayers. I sat beside her and listened to her every breath. Most were labored, but each extended her life. I held her hand."

Fidelis paused before adding, "Her hand was still warm. I wouldn't let go."

"I'm sorry sir, but I ask you to speak louder, so your testimony can be recorded." Admiral Timens wanted everyone to hear Fidelis.

"Yes, sir."

During his testimony, one last witness readied to appear. Only then would Timens reveal his true agenda. The Inquiry neared its conclusion.

[End of Chapter]

CHAPTER NINETEEN

Another Fourteen Minutes

Blocks from the Courthouse was Captain Astraea. She rested in the Campos Naval Hospital. Her injuries weren't healed. Every day was another painful reminder of her injuries. This one began no differently. Astraea's nurse helped her bathe and dress. After her morning meal, the Captain listened to media. When able, she accepted company. By afternoon, she prepared for her daily walk.

The Nurse entered Astraea's hospital room. She checked her patient's vitals. "Is it time for my walk?" asked the former *Vindictus* Captain.

"It is ma'am."

"Can we wait for Fidelis?"

"Ma'am, he won't be here this afternoon."

Astraea was disappointed. Every afternoon, Fidelis took her to the park. She remained in her wheelchair until they reached the garden. There she took a few steps with assistance from forearm crutches. Fidelis gently steadied her. Much of their time was spent together on park benches.

"But Fidelis always walks with me. Is he ill?"

The Nurse replied in a hushed voice, "No, ma'am. I brought your wheelchair."

"But I wish to walk. Where is Fidelis?"

"Someone said he was testifying in the Navy Courthouse across town."

The Captain's Aide entered. It was Lieutenant Commander Negatio.

"Is that you, Mr. Negatio?"

"It is ma'am." She reached for him and asked, "May I hold your hand, sir?" He knew it was time for her afternoon walk.

Of course, "May I walk with you today, Captain Astraea?"

"I would be honored, sir. They tell me Fidelis is testifying at the Naval Courthouse? Why?"

"I can't say, ma'am."

"What is it about?"

Negatio hesitated. He didn't want to upset his former Captain, but he wouldn't hide truth from her. "It's about the Battle of Sector 12, ma'am."

"But Fidelis knows nothing about that. It's odd they would seek his testimony. Please take me there. I wish to hear him."

Astraea turned to her Nurse. She needed a change of clothes. "I would like to wear my dress white uniform to honor my crew."

"I'm sorry, ma'am, we don't have your dress whites. We do have dress blues."

Negatio had anticipated the Captain's preference. "Captain, I brought your dress whites." He handed them to the nurse and stepped outside momentarily. Within minutes, the patient was sharply attired. The Commander was beckoned back to the room. He was stunned by Astraea's appearance. "Ma'am, you're beautiful!"

The Lieutenant Commander helped Astraea into her wheelchair. He carried her arm crutches in a standard issue carry-bag. Negatio helped his former Captain walk with aid of crutches to awaiting medical transportation.

Astraea jested, "Mr. Negatio, did you also bring white lace for me?"

"No, ma'am. There aren't any Phasmii there," he quipped, "but I did bring your helmet and battle blade, just in case." Astraea laughed. "Thank you, sir!"

The two were transported from the hospital grounds to Campos Naval Courthouse. Astraea's nurse accompanied them. Captain Astraea was seated in her wheelchair as Mr. Negatio steered her into the building. As they neared the main courtroom, the two overheard loud voices on the other side of large, wood-paneled doors. Inside was the grand courtroom. Admirals Pavidas and Pullum suddenly burst through the doors and scurried away. Pullum's robe was nearly caught in the closing doors.

"What was that, Mr. Negatio?" Astraea asked.

"It was nothing, nobody," ma'am.

The Sergeant of Arms exited the courtroom. He watched the two Admirals scurry away. The Sergeant approached Astraea and Mr. Negatio. He saluted her. "Good afternoon, ma'am. Vice Admiral Timens shall receive you soon."

"Our Captain Timens is a Vice Admiral? How wonderful! Why is he expecting me?"

Negatio quietly informed her. "Captain, while you dressed, I advised the Vice Admiral we were coming."

After clearing his throat, the Sergeant of Arms reported, "Captain Astraea, Vice Admiral Timens requests that you wait here while Mr. Fidelis testifies."

"I want to hear Fidelis."

"Yes, ma'am. His testimony will be transmitted to this room. The Vice Admiral asks that you wait here until the Councilman has testified. Afterward, I will take you into the courtroom."

"Sergeant, Mr. Negatio shall escort me. I will not enter in this wheelchair."

The Captain's Nurse urged her to reconsider. Astraea insisted she wouldn't be wheeled into the hearing. Negatio had prepared for that eventuality. "I was afraid you would say that ma'am, so I brought your arm crutches."

"I won't wear those, either. The people inside are the loved ones and families of our crew. They don't want to see me as a cripple. Please, let me walk without crutches."

Her Nurse cautioned, "Ma'am, that's not possible! Your doctors have ordered otherwise."

"Mr. Negatio, are any of my doctors here?"

"No, Ma'am."

"Do you order me to enter the Inquiry on crutches?"

"I can't order such a thing, but please use them."

"Please, sir. Assist me instead."

Mr. Negatio bit his lower lip. He knew Captain Astraea would try to walk alone if he declined. So, he reluctantly accepted. "I will walk with you, ma'am, but please take my arm."

"I am honored, sir."

Vice Admiral Timens spoke loudly enough for Astraea to hear outside the courtroom. He asked, "Please, Councilman Fidelis. Tell us of Captain Astraea."

Fidelis didn't know Astraea was there. He had no idea she was listening."

"We won't discuss the Battle at Sector 12. Please tell us instead how you met the Captain?"

Everyone in the gallery remained quiet. The Inquiry was supposed to be about the Battle. Yet, they all wanted to know more about Astraea.

"I was having lunch at 'Doggy Joe's.' It's a hot dog restaurant in my ward. Astraea was having lunch with a group of young officers. I thought military people were stiff and disciplined, but they were having a great time. All of them were laughing, joking, or chatting. She caught my eye.'

"Admiral," he continued, "you couldn't help but notice Astraea! She was amazingly beautiful. I was captivated by her every laugh and smile. I tried not to gawk at her, but I couldn't look away.'

"She and her friends finished lunch. They paid for their meals and left. Astraea remained. She noticed a needy family. Their dress suggested they couldn't afford their meal, much less ice cream for the children."

Timens smiled kindly. "Mr. Fidelis, I remember how observant she was."

"Astraea paid the cashier for their meal and ice cream. I heard her tell the cashier not to tell them she covered their tab. That moved me. Before I could turn my head, Astraea approached. I was embarrassed that she saw me watching.'

"As she passed by me, an ice drink slipped off her tray. It spilled all over my trousers! I expected a cold shoulder, but not an icy lap!'

"Astraea was apologetic. 'I'm so sorry,' she said. When I looked up, I met her glistening, green eyes. I couldn't avert her gaze. There was an incredible connection between us. I tried to deny it. Certainly, she couldn't feel the same thing.'

"Astraea didn't appear regretful. Instead, she seemed rather pleased with herself! I didn't care! It was worth the price of her smile. Astraea mopped my table with napkins. I told her not to worry. I would take care of it. I thanked her for her kindness."

"What happened then, Mr. Fidelis?"

"She sat at my table! I don't remember what she said. I didn't care.

Then she returned to my booth with two more iced drinks. I thought she was going to spill both on me, but she sat across the table and laughed. We began talking. It was about something silly. I don't remember what. We had ice cream together. Then, we left and went to the Park. We spent the rest of the afternoon together."

Timens interrupted. "Did you miss work, sir?"

"It was a holiday, Admiral."

"Indeed, it was. Proceed."

"We had dinner together. There was nothing about Astraea I couldn't love. I wished the day would never end."

"But it did, sir?"

"Of course, I never expected to see her again. I was grateful for having that single day. I expected her Naval duties and friends would carry her far away."

"But you did see her again?'

"Astraea called me the next day. She said she was stuck at Doggy Joes with an extra iced tea. She asked if I could join her, so she wouldn't have to drink it. I figured I was being pranked. I went anyway. I wanted to see her."

"Were you pranked, Councilman?"

"No, she waited for me with two ice teas. As I sat at the table, she smiled and acted like she was going to spill one on me. I couldn't help but laugh. We chatted. I didn't want to leave without her."

"Was that a holiday, sir?"

"No, sir, but the day became a week. The week became forever. I think I've always loved Astraea, even before I met her. I always will."

"Did you know she was one of my officers?"

"So, I've been told."

"Did she have kind things to say about me, sir?"

The Gallery chuckled. Fidelis feigned discomfort. He straightened his tie.

"What say you, Mr. Fidelis?"

"Astraea has kind words for everyone, as you must know."

"I do, sir. However, Admirals Pullam and Pavidas raised concerns about her judgment. May we address their concerns, Councilman?"

Many in the gallery murmured. What was Timens trying to do? Why was he questioning Captain Astraea's judgment?

With crossed arms, Fidelis replied. "I don't know the other Admirals' concerns, sir."

"Mr. Fidelis, they questioned the Captain's judgment."

"About what, sir?"

"They questioned whether she should have captained the *Vindictus* in battle."

"Did she have a choice, sir?"

"They suggest she misjudged the situation."

For the first time before the Inquiry, Fidelis was angered. Astraea lay in a hospital bed. She was gravely wounded and suffered constant pain. Her injuries were severe, life threatening. She nearly died in that battle.

The Vice Admiral continued. "Have you ever questioned her judgment, sir?"

The Councilman paused. The courtroom went silent. Everyone awaited his reply.

Quietly, he answered. "I have, Admiral."

There were gasps from the audience. Timens was surprised. "On what occasion?"

"I questioned her judgment when she invited me to drink iced tea with her. Astraea had many better choices."

"Then, I suggest that we test her judgment, Councilman."

"Sir?"

The Vice Admiral beckoned the Sergeant of Arms. "It's time," he directed. Two Marines in dress-blue uniforms opened the courtroom doors. They were Corporals Probus and Erectus. Chief Iratus stood by their side and saluted.

In view of the entire court was Captain Astraea. She was seated in a wheelchair. Surprised, Fidelis turned and saw her. He began to step toward the doorway, but Mr. Negatio waved him away. As one of her officers, it was Negatio's duty to escort the Captain.

"Captain Astraea," called the Vice Admiral. "Can you give us fourteen minutes?"

The phrase "fourteen minutes" spread through the gallery. Everyone knew what it meant. Captain Astraea asked the *Vindictus* crew to fight fourteen more minutes against the vicious Phasmii attack. The ship was dying, as were many of the crew. Most others were injured. All were exhausted. However, the crew gave her every minute. They saved the world from catastrophe.

"Captain Astraea, I ask only fourteen minutes, please."

Astraea arose slowly. With her Aide's assistance, the heralded *Vindictus* Captain stood and straightened her uniform. Lieutenant Commander Negatio attempted to fasten crutches to her forearms, but she waved them away. She would stand. Astraea was determined to *walk* into the Courtroom.

Negatio placed her arm through his. She held tightly. Her legs were weak. With every step, pain shot from her calves to her hips. Without the Lieutenant Commander's assistance, she would have fallen. They approached the Admiral together. Astraea asked, "Captain Timens, is that you?"

"It is, my dear. Somehow, I became a Vice Admiral. It surprised me, much as you surprised Mr. Fidelis."

Captain Astraea gently smiled. "Sir, may I approach?"

"You may, Captain."

Her walk was slow and unsteady. Every step worried the gallery. They feared she would drop to the floor. With Mr. Negatio's assistance, Astraea approached the Admiral's bench. Negatio then gave her arm to Fidelis. Lieutenant Commander Negatio apologized to the Councilman. Fidelis answered with gratitude.

Admiral Timens looked upon Astraea. Her gaze always drew his smile. He remembered how her green eyes turned blue on occasion. However, what he saw saddened him. Her eyes were colorless. Stunned, he lost his smile. Astraea was blind. Timens quickly recovered. He had duties to perform.

Astraea asked her former Captain, "May I touch your face, sir?" She reached upwards. The Vice Admiral leaned forward. He took her hand and placed it on his cheek. She gently felt his face. "It's good to see you again, sir. We worried for you."

Admiral Timens answered with difficulty. "Captain, the Navy called you to duty when the Phasmii attacked. We took you from your family and friends. We took you from your wedding and placed you in command of a dreadnaught. We sent you to war. When defeat was imminent aboard the *Vindictus*, you asked your brave crew for fourteen minutes, which many of them didn't have. I ask that you now give me fourteen minutes of your time."

Astraea did not understand what the Admiral asked. She would not deny his request. "Aye, sir. I shall do my best."

Timens turned to the Councilman. "Mr. Fidelis, you questioned the Captain's judgment. Let us test it again."

"Sir?"

"Answer me, Mr. Fidelis. Why would Astraea have you as her husband?"

The gallery whispered in confusion.

“Admiral,” Fidelis answered, “I cannot speak her feelings.”

“Then, speak your own, sir. Why have you love for her?”

The Councilman paused. He looked at Astraea’s face. Red streaks crossed her cheek from radiation burns. Her eyes were blank. Yet, Fidelis easily found her beauty. He gently took her hand.

“Mr. Fidelis, please answer the Court’s question. Captain Astraea awaits your reply.”

“Admiral, I cannot help but love her. Astraea brings cheer to sadness. She brings hope where there is despair. She makes important those who are not. She gives purpose to us all.”

“Is that your answer, Mr. Fidelis?”

“For each day with her, I thank our gods.”

Admiral Timens continued, “Let us now hear from Captain Astraea.” The gallery quieted. Everyone wanted to hear the legendary hero. She had made no public appearances since the Battle of Sector 12. Many had assumed she perished. Yet, she would soon answer questions from the Captain whom she had replaced.

"Captain Astraea, our fourteen minutes shall soon end. Please bear with me. It is important.'

"This Inquiry makes the following findings of fact; First, it is determined by this Inquiry that Councilman Fidelis loves you, and he is devoted to you. Second, it is found as fact that he shall one day become a great man – with your guidance. Third, Captain, let the record show your love for Mr. Fidelis proves your judgment is sound. Have you any objection to the Findings, ma'am?"

Astraea could not speak. She struggled with her emotions, as well as pain. Before she could gather herself, Admiral Timens beckoned the Sergeant of Arms. "Please bring the others."

Neither Astraea, nor Fidelis knew what was happening. The gallery turned and watched as the Captain's family entered the courtroom. Then, the Councilman's family entered. Their friends and comrades followed. Many in the gallery stood and offered their seats, so the couple's loved ones could honor them. The courtroom was packed with well-wishers. Astraea turned when she heard their voices. Fidelis inquired, "What is this, sir?"

The Admiral stood and looked upon the gathering. He pronounced. "Captain Astraea, when I learned of this Inquiry, I invited your loved ones here. This Inquiry asks, Ma'am, whether we may return your wedding to you?"

The Captain turned towards the gallery. Although blind, she saw them with her heart. She felt their presence. Astraea heard her sisters and mother sobbing. Then, some in the gallery began tapping their feet. They continued until almost everyone in the courtroom did the same.

There were murmurs of Astraea's name. Then chanting began, "We want a wedding." Astraea broke into a smile. Tears streamed down her cheeks.

"They're here!" exclaimed Fidelis. "It's our wedding day returned."

"Captain Astraea, you stand here, sworn to the gods. You've heard the Councilman's testimony. I ask you, ma'am, is it true? Have you love for Fidelis?

"I have, sir." She held Fidelis tightly and continued. "He needs not hear it, Admiral. He knows it. We know our love for one another."

"Never silence such love, Captain. Speak it freely."

The Captain's knees wobbled slightly. She stood too long without crutches. Her legs trembled. Yet, she stood with her head held high and shoulders swept back. Fidelis knelt and bowed his head.

Astraea ran her fingers through her beloved's hair and spoke. "When first we met, I spilled cold drink upon you, sir. Yet, you returned to me the following day, and the day afterward, and every day. When called to duty, I left you at the Templum with a promise that I would return. Only providence kept that promise from being broken. Only by the grace of gods do I stand here with you. At last, my beloved, I tell you before witnesses what you already know. Your laughter is mine, as mine is yours. I treasure your smile. It gives me comfort. I know it well. Although I no longer see it, I do feel it. Even in darkness, I know you, sir. I treasure your love. Please know mine. I promise to love you always.'

"And this I also promise," she continued. "I shall be your mystery. You shall always wonder whether my drink spills upon you by accident or purpose."

There were both chuckles and sobs from the gallery.

"I love you, Fidelis. Thank you for loving me."

The Admiral cleared his throat. "Please rise Mr. Fidelis." He turned to the Captain. "Do you, Alexia Arabella Astraea, take this man, Apius Caecus Fidelis, to be your lawfully wedded husband?"

"I do, sir."

"Mr. Fidelis, you stand before me as the most fortunate man alive. Do you, take this woman, Alexia Arabella Astraea, to be your lawfully wedded wife?

"I do with all my heart, forever."

The Vice Admiral looked for Fidelis' attendant, but Juvo wasn't there. No one brought Astraea's betrothal pin. Tradition required it to be placed above the bride's heart. Before the fourteenth minute, Juvo burst into the courtroom. "He called out, "I got it. I got it!" He held up high a betrothal pin. However, it wasn't the pin to give

Astraea. Instead, it was the pin meant for Fidelis. Juvo brought the same pin, which she had held in her hand aboard the shattered *Vindictus*. He retrieved it from her hospital room.

There was no bridal pin for Fidelis to give Astraea. It was across town in his apartment. Time was running out. Without a pin, the wedding couldn't proceed. Fidelis didn't know what to do. Astraea was unaware of the problem. She stood blindly awaiting Fidelis to pin her heart.

The Admiral wouldn't allow the moment to pass. He removed from his jacket one of several meritus pins. He had worn it above his heart for several years. It was Timens' most treasured decoration. Fidelis appreciated the gesture, but the wedding required a pin which Astraea had given. A matrimonial pin represented happiness returned.

"Mr. Fidelis," beckoned the Admiral, "I have such pin."

The gallery murmured. How could Astraea have pinned Admiral Timens? Had there been a scandalous relationship between them?

Timens answered, "I received this meritus pin shortly after receiving command of the *Vindictus.* It was given to me by my crew, among whom was your bride. It is very precious to me. I've worn it proudly. Please return it to her.

Admiral Timens asked the bride, "Alexia Arabella Astraea, will you accept this pin?"

"It honors me, sir."

He handed it to Fidelis. "Place it above her heart, sir."

Fidelis rose. With trembling hands, he placed the pin above Astraea's heart.

Timens concluded, "You have given me fourteen minutes, Captain. By the powers vested in me as Chairman of these Proceedings, in the presence of your families, friends, and the gods themselves, I pronounce you, Alexia Arabella Astraea, and you, Apius Caecus Fidelis, as husband and wife. You shall live this day, and every next day, together in matrimony. Your lives shall be shared as one. Amen."

The gallery replied in unison, “Amen.”

The Admiral looked at his watch. He continued, “Mr. Fidelis, our fourteen minutes are ended. It is time for you to kiss the bride.”

And so, he did. He held Astraea gently. Their lips met. It was the kiss she promised before her call to war. One of her sisters giggled. The other wept. Emotions swept the gallery. Fourteen minutes had passed. Yet, they would forever continue.

Timens was satisfied. He had completed his agendum. The Vice Admiral had defended the memories and reputations of his crew. He returned to Astraea the bridal day the war had denied her. He adjourned the Inquiry.

However, it was not a “happily ever after” moment. Destiny would call upon Astraea and Fidelis again. Ominous days awaited.

[End of Chapter]

CHAPTER TWENTY

Captain Astraea Meets Prime Minister Morus

Three months passed. Everyone watched the skies. The Phasmii menace had ended. However, another catastrophic threat revealed itself. An asteroid known as Rogares grew a fiery tail. It fractured into a rocky, icy comet and changed its elliptical orbit. Its distance from Gemina narrowed each time it passed. After thirty-two more orbits, Rogares would crash into the planet. The collision would be cataclysmic. The world had less than four years to live. Each day brought apocalypse nearer.

Citizens expected their military to divert or destroy Rogares. They had confidence in the Navy that defeated the Phasmii. Yet, no ships were sent to ward off Rogares, despite growing public demands.

The government constantly assured the world there was no reason to worry about the rogue comet. At first, the Administration explained the fleet was undergoing repairs from the Phasmii attack. Later, it explained to a frightened public that the fleet couldn't attack Rogares until the comet reached an apex far away from Gemina. That never happened. The government gave technical excuses. The public grew impatient. The Legislature responded by authorizing the construction of seventy-two technologically advanced warships.

Prime Minister Morus was aware of the citizenry's growing concerns. He knew public trust in his government was eroding. His poll numbers decreased each week. Something had to be done.

Captain Astraea asked for an audience. Morus gladly accepted. He believed their meeting would boost confidence in his government. So, it was announced publicly. The Prime Minister promised to discuss ideas about promoting government bonds or improving benefits for veterans and survivors of the Phasmii war. He knew those issues were close to the Captain's heart. However, Morus also intended to address the Rogares issue.

Captain Astraea had been severely injured by the Phasmii attack. However, she wouldn't seek public sympathy, which detracted from the plight of her crewmates and their families. Few knew she was blind, or that she still struggled with life-threatening wounds from the Battle of Sector 12.

Meeting the Prime Minister would be difficult for her. She could only walk short distances without her cane or walker. Standing and sitting were limited to short periods. Yet, she would not decline the opportunity. It was an opportunity to advance the welfare of blind or disabled veterans, as well as families of deceased servicemen.

Astraea was the world's most celebrated hero. Her only public appearance since the Phasmii attack had been the Naval Inquiry. It concluded with her famous wedding to Fidelis, after which other couples planned fourteen-minute ceremonies. She had captured the public's heart. Geminum wanted to see Captain Astraea, hear her speak, touch her.

Hundreds of well-wishers gathered by the roadside to watch the Captain's motorcade. They waved to her enthusiastically. Parents held their small children high to see the famous heroine. Many cheered her name. Once her motorcade arrived at the Government Ministry, the

Captain was helped from her limousine. She smiled and waved to the waiting crowd.

Her visit to the Ministry was an extraordinary opportunity for journalists, commonly referred to as "journos." They understood the public's hunger for news about the legendary Captain. Dozens stalked the Captain's naval motorcade to cover her meeting with the Prime Minister. Others awaited Astraea at the Ministry where they would harry her with witty questions.

The Captain's Naval Escort was Cadet Acquila Bellona. Many questioned why Miss Bellona was assigned to such an important duty. She was a midshipman at the Naval Academy. Bellona was an orphan, or "papilli," as they were called. Another papilli, Senator Orba, had recommended her to serve as Astraea's escort. The Senator was a confidante of Prime Minister Morus.

Both women wore dress blue uniforms. Each sheathed a ceremonial dagger on her left thigh, as was customary. The Captain proudly held her plumed Cuirassier style helmet on her right side per protocol. Bellona wore hers, as she accompanied Astraea to the Ministry. Her helmet was strapped below her chin and properly tilted forward, which symbolized respect and honor.

Their limousine parked in front of the Ministry. Cadet Bellona helped the Captain from the vehicle. They walked together towards the Ministry entrance. Journos, who were not known for courtesy or respect, rushed Astraea. They intended to block her path and force her into a press conference.

Two raced hectically towards the blind hero. The remaining gaggle followed. Their number grew. The rabble rushed Astraea. She was about to be trampled.

She paused after hearing the approaching journo horde. Cadet Bellona braced the Captain. As she did, the Cadet dropped her dress helmet. Such clumsiness was disgraceful. Among military circles, it drew derision, even punishment.

The journos madly hurried towards Astraea. The leader of the news hounds was a leg length away when Cadet Bellona knelt to retrieve her helmet. Her leg inadvertently brushed the journo. He stumbled, fell, and crashed onto the pavement. Another one tripped over him, as did those who followed. They piled atop one another into a scrum.

That cleared the legendary Captain's path. Cadet Bellona guided Astraea forward. As they ascended the Ministry steps, Cadet Bellona quietly called each step to the Captain. That helped her ascend gracefully. Anyone who watched would have assumed Astraea needed no assistance.

Atop the portico, Astraea spoke quietly with her escort. A casual observer may assume she reprimanded her escort for dropping her ceremonial helmet. That was incorrect. "Cadet Ensign Bellona?" the Captain asked. Her voice was hushed so others could not hear.

"Yes, ma'am."

"I may be blind, but I know what you did."

"I'm sorry, Captain. I lost my helmet for a moment."

Astraea laughed. "Miss Bellona, you never lost your headgear. You dropped it on purpose. Didn't you?"

The cadet paused. Reluctantly, she answered, "Aye, ma'am."

"And you executed a leg sweep on the leading journo? Am I wrong?"

"You knew that, Ma'am?"

"Which sweep did you use?"

"The forward leg sweep, Captain."

"Job well done, Cadet. There's something about you, Miss Bellona. They say you're a papilli. Somehow, I believe destiny is your mother. One day we will learn more about your father."

Two Marines opened the Ministry's tall, walnut doors for Astraea and her escort. Each of them saluted Captain Astraea. Although blind, she returned their salutations. Her visit proceeded with dignity as the journos untangled themselves from one another outside.

Standing by the Prime Minister's doorway was his Sergeant of Arms. He saluted Astraea and announced. "Prime Minister, the Honorable Captain Alexia Arabella

Astraea of the Gemina Galactic Navy requests an audience."

A distinguished, gray-haired gentleman rose from his desk. He straightened his gray, worsted wool jacket and motioned forward. "Please, enter Captain." Astraea and Cadet Bellona joined Morus. The midshipman stood at attention behind the legendary heroine.

The Prime Minister was genuinely humbled by Astraea's presence. Unaware of her blindness, he smiled and offered her his hand. When he looked into her eyes, Morus realized she was blind. He was saddened and felt foolish. The Prime Minister began to withdraw his hand. However, the Captain held hers out to him, which he gently grasped. "Thank you for coming, Captain Astraea. I'm honored by your presence." He added, "It's not every day one is graced with the company of greatness."

Astraea nodded her head slightly and joked. "That's what I tell my husband every morning. Yet, for some reason, my dear Fidelis thinks I'm honored to be in his company!"

Morus laughed. "Please Captain, have a seat." Astraea sat across the Prime Minister's famous desk. He resumed his place. "Captain, I once asked your husband to join my Administration. He declined."

"That doesn't surprise me, sir. He finds enough problems to solve in a single City District, much less a Province. My poor husband couldn't get to our wedding without servicing his constituents. He seeks no higher office than City Councilman. Fidelis tells me he's not suited for grander duties."

"Captain, Fidelis will learn one day that public servants don't choose their duties. Their duties choose them."

The Prime Minister shared a moment with Astraea. "I remember that terrible day when you and your brave crew fought in Sector 12. I prayed for your protection. Some among us wept as the *Vindictus* drifted helplessly, and the enemy sent their largest ships to attack you. All seemed lost. Then, suddenly, the Phasmii were destroyed! Admiral Timens told me our secret weapon won the battle. I knew of no such thing, so I asked what was it? He replied, 'It was Captain Astraea, sir.' Now, sitting across my desk is that great weapon. Captain, thank you for coming. Now, please tell me what we may discuss together?"

"Thank you, sir. Cadet Ensign Bellona, you may leave us alone. I shall call upon you after the Prime Minister and I have concluded our conversation."

"Yes, ma'am," answered the raven-haired cadet. She turned sharply and exited the room. The Prime Minister signaled for the Sergeant of Arms to leave, as well. That gave Astraea and Morus the privacy required for their candid conversation.

The Captain began. "Prime Minister, may we speak about the growing concern among our people?"

"What is that concern?"

"Comet Rogares."

"Yes, it haunts me."

"The public is anxious, sir. The thing races towards us on a collision course. Their angst grows daily. It will worsen until Rogares no longer dooms us."

"I understand."

"Sir, panic is not a visitor we should welcome to our homes."

"No, of course not. Have you any proposals, Captain?"

Astraea replied. "Yes, the government should announce a mission to divert Rogares. That would ease the people's anxiety and restore calm. We should also rebuild scientific installations on Semotus. From there, we could better monitor the Sectors nearest us."

"What mission against Rogares would you recommend?"

"We require but one ship, sir. I suggest we send the *Lioness.* Commander Clades commands her. He is our best Captain, and she is our most powerful warship. The *Lioness* is a G Class Destroyer with ample speed and firepower to divert the Comet."

"Do you believe your plan will bring calm to the world?"

"Yes, sir. Public confidence in the government will soar when they learn that the *Lioness* defends us. Everyone knows she fought off sixty-four Phasmii warships before they engaged the *Vindictus*."

Reluctantly, the Prime Minister paused. "I'm sorry, Captain. I cannot grant either of your requests."

His response surprised Astraea. "May I ask why not, sir?"

A pained look crossed the Prime Minister's face. Astraea couldn't see it, but she heard it in his voice. Reluctantly, the Prime Minister began his answer, "I've thought much about the problem. What you say is true. The public grows anxious. Something must be done about Rogares. However, I can't accept your proposals."

"But with all due respect Prime Minister, why not?"

"What you propose promises disaster."

"Why do you say that, sir?"

"I remember the terrible day when Commander Clades and his squadron brought you and your crew home from Sector 12. I questioned why the Phasmii attacked us with over three hundred warships? What provoked them?"

“Can you explain their actions, sir?”

“I concluded that I had provoked them by doing what you now suggest. I was the one who authorized Outpost Semotus. I sent warships into a galactic neighborhood. Captain, we angered at least one neighbor. Should we risk offending another? More are out there. I’m certain of it.”

“Prime Minister, we had fifteen warships and a defenseless outpost. How could that have provoked an armed response by three hundred and twenty-eight warships? We were no threat to them.”

“Yet, you destroyed them all with one ship. How will they react to another Gemina warship voyaging into their space?”

“Prime Minister, we speak privately within four walls. No one else hears our words. I tell you, sir, the Phasmii did not attack in self-defense. They attacked with hate and rage. For that very reason, I pray they believe the *Vindictus* destroyed their whole armada. Then, when they see the *Lioness*, the Phasmii will flee in terror. They won’t dare risk combat against us, nor will anyone else.”

Morus understood her argument. He had already considered and rejected it. The Prime Minister explained, "If any of our neighbors believe we are a threat, we invite another armed response. How would we defend ourselves? We now have only seven ships, mostly frigates. No longer do we have the mighty *Vindictus*, nor her valiant Captain and crew."

"Sir, I speak frankly. You did not cause the attack on Semotus, nor the Battles of Sectors 12 and 16. It may comfort you to believe you had such power. You may want to believe you could prevent such attacks. However, you had no such power then, nor do you have it now. The Phasmii, or whoever, will attack us for reasons of their own, or no reason at all. You cannot decide their actions for them."

Morus considered her words. What she said had reason, but it did not sway him. However, he didn't want to offend Astraea with disagreement. More than anyone, she had his respect. He paused.

"Sir," the Captain continued, "Good neighbors enjoy no special protection from malice. Please dispatch the *Lioness* to divert Rogares. Until then, Comet Rogares will curse us. Geminum will live in daily fear. Please, spare them that burden."

"Captain Astraea, I hope you understand I can't send the *Lioness* or any other warship across the galaxy."

"Prime Minister, the people have spoken on the issue. The Legislature authorized construction of a new fleet. They're building seventy-two new ships. All of them will be faster, quicker, and more powerful than anything that fought in Sector 12. Their technology is far more advanced than can be imagined."

"What you say is true, Captain. Those ships are being built as we speak. I can't veto their construction. However, it is my executive prerogative to determine their use. We will not militarize space, no matter how many ships are constructed." Morus paused uncomfortably. He had wanted a pleasant conversation. He never wished to quarrel with Captain Astraea.

"Prime Minister, it's time we part," she said. "Thank you for listening to me. I ask you to consider my suggestions. We must act soon. If not, it won't be the Phasmii that destroy us. We shall be our own undoing."

Astraea called upon Cadet Bellona. The midshipman entered the chambers and assisted the Captain. Morus also rose. "Captain Astraea, thank you for coming. I shall consider your recommendations."

"Thank you, sir. I ask nothing more."

With a nod and handshake, Astraea and her escort left his office.

Morus leaned back in his chair and pondered the Captain's thoughts. There was merit in what she suggested. However, he would not dispatch the *Lioness* to divert Rogares. He was determined to avoid another war. There was another alternative.

Morus opened his comvid. He addressed his Aide. "This is the Prime Minister. I need you to do two things. *First*, reassign Commander Clades from the *Lioness*. Yes, I understand he is her Captain. That's why I want him transferred to the Academy. Make him an instructor there. No, it can't wait. I want it done, now. Assign someone else to command that ship. I don't care who. Just do it.

Second, and this must be done immediately, activate Commander Scitus. Yes, I know he's retired. Call him out of retirement. I need Scitus for an important mission. No, I don't want him to command the *Lioness*. Give him a cutter. Better yet, activate the new SAP Cutter. Get her ready. I want her armed. Give the ship a Level Three Cannon. Of course, cutters aren't designed to carry such weapons. Don't you think I know that? That's why I chose it to be his ship. This is a matter of urgency. It's top secret. Do you understand? Very well. Thank you, Colonel. I'm depending on you."

Morus closed his comvid. He wouldn't give Scitus a warship to divert Rogares. Instead, he assigned him to a small, lightly armored reconnaissance vessel. Morus believed that wouldn't provoke a military response. With a successful mission, the world would again live free from anxiety and dread. At least, that was his plan.

He closed his eyes and prayed.

[End of Chapter]

CHAPTER TWENTY-ONE

Blood Alley – Prelude to Comet Day

Ensign Amator piloted his jet-black cutter alongside Convoy X-Cargo 9. He led a squadron of six Mark 10 Cutters to escort the convoy through an ugly, galactic gloaming known as "Blood Alley." Many ships had been attacked there. Scores were destroyed. Hundreds of crewmen and passengers had been killed or were missing and presumed dead.

The convoy welcomed the sight of Amator's "Dragon Squadron" and his menacing Cutter. It appeared eager and ready for any fight. Painted on the cutter's nose was a fierce, toothy shark snarl with angry, glaring eyes. However, his Cutter wasn't built for combat. It was a lightly armored reconnaissance ship forced into that role.

Merchantmen called Amator's jet black Cutter the *Raven.* The Ensign modified it for escort duty. He replaced its standard-issue automatic gun with a powerful Level Two Cannon.

His wingman, Ensign Socious, modified his Cutter similarly. His ship also wore a snarling shark face. It was painted raging red, instead of traditional Navy gray and blue. His fierce-looking Cutter was named *Firebird.*

It, too, carried a Level Two Cannon.

The Captain of the lead merchant ship hailed Amator. "It's good to see you, Dragon Squadron." The convoy ships signaled welcomes to their escorts by flashing docking lights on and off.

"Convoy X-Cargo 9, this is Ensign Amator. We will escort you for the next 90 light minutes. Please form a close column formation. *GSS Mercator*, you may resume as pacemaker."

"Aye, Ensign Amator. I don't mind telling you we're scared. This is the same patch of black hell where X-Cargo 7 was attacked three days ago. Only one of those ships survived. The others were blown apart, except the *Zong*. She was grappled and towed away into the darkness. Only the gods know what happened to her. We don't know who attacked them, *Raven*. Nobody thinks they're pirates. Some fear they're Phasmii."

"Fly straight, Captain. We'll get you home."

"Raven?"

"Yes, Captain."

"There are thousands of bends in space throughout this Sector. There could be warships hiding in them."

Amator replied, "Keep a good lookout, sir. We'll do the same."

The Ensign searched the Sector for anomalies. His communications equipment crackled. It was his father, Admiral Pullum.

"Son, this is your father. I need you to transfer home. Your mother cries herself to sleep every night. She's frightened. You're in constant danger. You've completed three tours. It's time to come home, Amator. We need you here."

Pullum's request was interrupted by another transmission. "Ensign Amator, this is Dragon 3. I'm picking up unusually high baryon readings approaching. I will continue monitoring."

Amator checked his equipment. “I’m busy here, Dad. I’ve already told you. I’m not returning to Gemina until you clear Gemma of that bogus conviction you put on her.”

“You’re right, son. I’ve arranged to set aside the conviction. She won’t be a fugitive any longer. Just come home, please.”

“Is that true?”

It was. The Admiral arranged for Gemma’s conviction to be set aside. Gemma would be cleared of all charges. However, what he did not tell Amator was that Gemma was presumed to have been killed-in-action during the Phasmii attack on Outpost Semotus. Setting aside her conviction was a phony gesture to lure Amator home.

“By the time you come planet side,” Pullum assured Amator, “your friend Gemma will have been cleared. I promise you that.”

“Do I have your word on that?”

“You do, son.”

“Alright, tell Mom I’m coming home. I’ll sign the transfer paperwork after I finish this sortie.”

"Thank you, son. We're both proud of you."

"Tell Mom I love her."

"I will."

After a pause, Amator continued. "Oh, Dad?"

"Yes, son?"

"You know, I love you, too."

The transmission ended. Ensign Amator picked up anomalous readings. Unusually high levels of baryon particles approached the convoy at high speed. "Dragon Squadron, this is Amator. Be alert. I'm picking something up. It's headed towards us."

Moments later, a strange ship raced by the Cutters. It passed through the merchant ships.

"Amator," signaled Socious, "that's not a Phasmii ship, and I've never seen a pirate ship like it. That's something else."

"Take my wing position, *Firebird*. We're going after it."

"Aye, Ensign. I'm at your wing."

No sooner than they vectored away, disaster happened. The *GSS Mercator* violently burst into an exploding, blazing sun. It flashed brilliantly, then quickly disappeared into darkness. There was nothing left of her.

Amator couldn't know it, but *Mercator* had fallen victim to a gyrfalcon.

"This is *Firebird.* I detected cannon fire of undetermined type."

Amator cautioned the squadron. "Remember, our shields won't protect us from cannons. Stay out of their way.'

[Pause]

"I've got the bogey," reported Amator. "I'm in pursuit."

The Ensign didn't know he was pursuing a deadly Hosticci gyrfalcon. It was fast and powerful, although not as fast as a Mark 10 Cutter. He quickly overtook it.

"I'm on its six," reported Amator. I'm engaging the attacker. I have firing solution." Amator unleashed a cannon barrage upon the gyrfalcon. The enemy craft went dark and spun out of control into the void. Its broken plating flung wildly into the cosmos.

"*Firebird*, this is *Raven*, I'm tracking a second bogey. It's on your six. On my mark, begin a Thach-weave maneuver."

That maneuver leads the attacking vessel into the wingman's guns. The attacker would follow *Firebird* into Amator's gunsight to be destroyed by his Cutter's cannon fire. Amator acknowledged the communication.

"*Raven*," signaled *Firebird*. "Mark."

Firebird crossed in front of *Raven*. He was followed by the second gyrfalcon, which flew directly into *Raven's* weapon fire. The menacing black Cutter let loose a barrage of blistering cannon shots. The second gyrfalcon erupted into a blinding flash. Its fractured remnants tumbled into the expanse as darkness consumed the dead craft.

"Amator, there's another one of those things behind me. I think I can outrun it."

"Don't worry. I've got it, *Firebird.*" As it crossed in front of Ensign Amator's Raven, he obtained a firing solution. Amator pounded the craft with a fierce cannon barrage. The target twisted wildly before exploding into a bright white flash. Its scattered remnants were swept away by a burst of solar wind.

Raven, this is *Firebird*. I'm reading more baryon signatures. They're coming from our port side and from our starboard ... and from topside and below. They're everywhere. We're surrounded by them. There's no way out!"

Dozens of gyrfalcons targeted the two Galactic Cutters. *Raven* and *Firebird* dodged multiple cannons firing at them. Amator signaled Dragon Squadron to lead the convoy through the asteroid belt before its opening closed. "We'll keep them busy and catch up with you later."

Amator assured Socious. "There's always a way out, *Firebird*. Follow me."

Determined to fight his way through the attackers, Ensign Amator flew his Cutter directly at a surprised gyrfalcon. "Power up your thrusters, Socious. If anything gets in your way, hit it with everything you've got. Don't slow down."

"Aye, *Raven*. I'm with you."

"I'm going to blast the one ahead of us. When it disintegrates, there will be a gap in their formation. We'll race through it and go to maximum speed. They won't be able to catch us."

"Acknowledged."

With their cannons blazing, the two Cutters plowed through the swarm of gyrfalcons. They concentrated their fire power at an unfortunate gyrfalcon ahead of them - and blew it apart. *Firebird* quickly passed through the gap where the gyrfalcon had been. He powered away from his pursuers.

Amator wasn't as fortunate. He met determined opposition. Two gyrfalcons intercepted Amator's *Raven* as he raced toward an opening in their formation. The Ensign acquired target solution on the leader. As he began pounding it with cannon fire, the other gyrfalcon landed a single hit. The *Raven* immediately erupted into a fusion blast. Three Gyrfalcons passed so closely that they were burned. One was fried, another ruined. The *Raven's* scattered pieces swirled into blackness, leaving behind a glowing, blue space luster, which radiance continued for years. It served as a monument for the young pilot's bravery. The *Raven* was gone. Amator was no more.

[END OF CHAPTER]

CHAPTER TWENTY-TWO

Dawn of Comet Day

Forty minutes later. Admiral Pullum attempted to contact his son. "This is Admiral Pullum. Connect me to Ensign Amator. Unaware the Admiral was the Ensign's father, a flippant communications officer callously remarked, "The *Raven* is never more, never more." He later explained he wasn't aware the Ensign had been killed in action. He believed Amator had been transferred. His crude remark earned a court martial.

Pullum was emotionally wrecked. His greatest fear was realized. His son was dead. He couldn't fathom how to tell Amator's mother. The aggrieved father was in shock. He was dejected, yet furious. Adrenaline coursed through his veins. He promised that someone would pay for his son's death. The Admiral marched to the stockade. He intended to avenge Amator by murdering the infamous pirate leader, Captain Damnatus. The Admiral wanted to feel the man's bare neck in his hands. He would crush every bone in it. Pullum wanted to hear Damnatus helplessly gasp for air through a crushed windpipe.

At his command, guards dragged an unbathed, unshaven old man into an interrogation room. He wore sailor's clothing with a Captain's insignia. None of his attire matched. Over his heart were numerous medals made from buttons. While awaiting interrogation, the old man entertained himself. He sang ditties about drunken pirates.

Admiral Pullum entered the room alone. Damnatus knew something was amiss. He had been interrogated many times, but never by a single interrogator, and this one was an Admiral! Damnatus recognized the officer. His interrogator was the coward Pullum. He was one of the weasels who fled from duty when the Phasmii attacked. Afterwards, the yellow belly tried to foul the memories of brave heroes who fought in his place. That included Captain Astraea. How could such a despicable man dare interrogate Captain Damnatus? He would have none of it.

Pullum sat across a rectangular table from the pirate. "I'm Admiral Ixion Crassus Pullum," he began.

"Ah, I know you," replied Damnatus. "We calls you 'Chicken Legs." He brokered a crooked smile, revealing gnarled, yellow teeth.

Pullum leaned forwards. The pirate's wrinkled neck was within his reach. He imagined the feel of the old man's vertebra popping apart in his hands. He retorted, "And we call you a pirate, a criminal, a murderer."

"Does you, now?" Damnatus proudly brushed his Captain's insignia.

"Is Damnatus your real name?"

"It's real enough, Admiral Chicken Legs. Would you rather they called me, 'Captain Bob?"

"They say you were the Captain of the *Latro*?"

"I been one of her Captains. What's it to you?"

"You and your vicious bastards killed my son."

Damnatus didn't know what Pullum was talking about. He supposed it may have been true. His reply was, "You don't say?"

"I did say. He flew escort for a merchant convoy in Sector 6."

The pirate captain didn't remember any attacks in Sector 6, at least not recently. "Please tell me, sir. What was the boy's name?"

"Ensign Amator."

"Ahh, I knowed him! You say he's your son? I mean no offense, but are you certain you're the boy's father?"

"I'm certain that your cut-throats murdered him."

"So, it was your son who flew the black Cutter? He was a damn good pilot. We called him 'Raven,' because the boy brought us nothing but bad luck. He took apart three of our sloops in a single day. He was a good lad, a real fighter. I wished he were one of mine."

"You and your criminals murdered him, Damnatus."

The pirate Captain turned away. Although he held the Admiral in contempt, he couldn't endure the wrenching pain in another father's face. Damnatus had suffered similarly.

"Look at me, damn you." Pullum pounded the table with his fists. "I could kill you here and now. No one would say a thing about it."

Reluctantly, Damnatus answered Pullum. "They calls me 'Captain Damnatus,' and no, I did your boy no harm. It weren't no pirate that killed him. We stayed clear of the *Raven*. Whenever we seen the Black Cutter, we just let his ships pass. He was a killer. We wanted no part of him."

"You dare call yourself a Captain?"

"Aye, and we calls you 'Admiral' Chicken Legs.' Now, pardon me, sir. I have nothing else to say to you. You suffer enough already. As a father myself, I grieves with you."

"You're a murderer. I will have you and all your cutthroats spaced. Confess now, Damnatus, or I'll kill you with my bare hands."

"Please, sir. Make me speak no more. I doesn't want to hurt a grieving father, not even you."

"Confess, or I'll drag you to the space lock myself."

"That would be a terrible fate," replied the pirate Captain. "They would all know me as the man killed by 'Admiral Chicken Legs.' That dishonor is owed to someone else."

"Who is owed that honor, 'Captain?"

"You don't know? Is your Navy so blind they don't see what's happening in their own Sectors?"

"What do you mean?"

"Pirates don't like killing, sir. It's bad for business. We like robbing. If we has to kill, we leave no witnesses behind. As they say, 'dead men tell no tales."

"Stop blathering, Damnatus."

"Why would pirates destroy merchant vessels, Admiral? That makes no sense. They bring us treasures for the taking. Without merchantmen, we'd have no booty. No sir, we're not destroying the geese that lays us golden eggs. Are you hearing me?"

Pullam leaned back in his chair. He wondered whether there was truth to the pirate's story.

"Admiral Pullam," the old man warned. "Your vengeance lies elsewhere."

Damnatus paused to see if he had the Admiral's attention. The old pirate continued. "I wager you sir. Answer me this. Ain't it true them devils that attacks your convoys destroys all but one ship? Tell me, sir."

"Yes, they usually leave one ship. You should know that since your killers murdered them all."

In a hushed voice, Damnatus warned the Admiral. "Haven't you wondered why they leaves witnesses? They wants you to know they're out there. They wants you to come after them. When you do, they'll kill you all. Just wait and see."

"Your lies won't save you Damnatus. You will be tried and executed, maybe just executed."

The pirate Captain replied, "For that I thank you. At least my epithet won't read:

'HERE LIES CAPTAIN ATICA DAMNATUS,

A POOR SAILOR QUITE DEAD,

BECAUSE

ADMIRAL CHICKEN LEGS SHOT HIM IN HIS PURTY HEAD."

"Don't write your epithet too soon," sneered Pullum.

"Sir, let me be frank. I speak to you as one father to another. What killed *Raven* weren't one of us. It was something evil, horrible. I seen what they do." Damnatus reluctantly continued. "Admiral, I'll tell you something I ain't told none of your cocky little interrogators. We found the *Zong* last week."

Pullum was surprised. The *Zong* went missing during a recent attack on a merchant convoy. She was one of the finest civilian vessels ever built. "You found the *Zong*?" asked Pullum. "Or did you steal her? Where is she?"

"We found what was left of her in Sector 7. She was adrift in the expanse. There was no life from her, no communications, no warnings, nothing. We hailed her and got nothing. The ship still had a bit of power. Her lights fluttered, so we expected someone aboard. I thought she might still have valuables, so I chose to board her. I took my best men. They was armed with pulse weapons, just in case the ship's marines had a problem with us."

"Did the Marines welcome you, Damnatus?"

"Nobody welcomed us. There weren't no Marines. We found parts of them, nothing more. It was an ugly sight. There was blood, bone and flesh everywhere. Pieces of people lay here and there. Some stuck to the walls. It weren't no pirates, Admiral. It was evil incarnate, and it laid itself upon those poor souls. We found three still alive. They begged us for death. They wanted relief from their suffering and madness. They told us about monsters them things brought aboard their ship. They called them '*metuses*.' Hades himself would cringe at the sight of them horrible things."

"Are you saying the Phasmii have returned?"

"No, sir. We detected baryon particles."

"So, what?"

"Wherever you find baryon particles, run Admiral. Them hateful things reek of baryon. Wherever is their stench, death awaits. Don't await them, sir. Just flee as fast as you can. Even then you may not escape. I know. Six of my sloops detected baryon and run away as quick as they could, but they was chased down. Each one was destroyed, at least the lucky ones. Them horrors grappled away one of my ships, just like they did the *Zong*. So, I sent three frigates against them."

"You expect me to believe you?"

"Believe what you will. The devils replied by sending dozens of them cursed 'gyrfalcons' after my ships. That's what they call their damned hounds from hades. They're fast, powerful, and near impervious. They hit, run, obliterate, and drag their victims away for their own amusement. That's what your boy faced, Admiral. Your fancy-pants Admirals sent him up there to die. The lad had no chance."

Pullum's angst worsened. Killing Damnatus would accomplish nothing. His son's murderers awaited him in Sector Six.

"Admiral, we heard tales of them wicked things for months now. Miners beyond the asteroids told us about them. They call them "Hosticci," because they was spawned in the Hosticcus System. I hear they like killing, Admiral.

It's a game they enjoy, so they make it last so long as they can. Them things draw every hope, every pain, every tear and scream from you before they let you die."

Damnatus paused when he saw tears escape the Admiral's watery, bloodshot eyes.

"I tell you, sir. I bet your boy give a good account of hisself, but he had no chance. As good a pilot as was the Raven, them little Cutters ain't no match against a gyrfalcon, much less swarms of the cursed things."

Pullum turned to leave.

"Admiral, you got steel to you. I never knowed it before, but you does have it. Mr. Pullam, sir. Don't go up there alone looking for vengeance. They'll kill you for sure. They'll torture away your soul. You'll gladly embrace death before they're done with you. Your boy wouldn't want you there."

The Admiral concluded his interrogation. Damnatus was dragged back to his cell. Vengeance wasn't found in the interrogation room. Nothing could mend Pullum's broken heart.

Pullum departed the stockade. He was exhausted, emotionally shattered. The Admiral wandered aimlessly towards the Ministry. Thousands of angry protestors surrounded it. Many chanted, "Morus ain't for us." Others yelled "It ain't gravy. Send our Navy." Protest signs reflected those slogans and more. One read, "Vomit the Comet." A much smaller group of counter-protestors held signs that read, "Give peace a chance." They shouted, "No more war!" Journos were everywhere. The Admiral pushed his way through them all. A journo from Channel One stepped in his way to get an interview and was knocked to the ground. Pullum pushed another aside and climbed the building's marble steps. Two marine guards saluted him, but he never saw them. Dazed, he continued onward, oblivious to reality.

There were loud voices ahead, inside the Prime Minister's office. Three union leaders argued with Prime Minister Morus. They wanted Gemina's military to defend the planet against commerce raiders and the rogue comet.

No one expected Admiral Pullum or his startling revelations.

Morus was about to address Miss Vectura. She was President of the powerful "United Transport Workers Union." Before the Prime Minister could utter a word, his office door burst open. In rushed an enervated Pullam. His sorrowful eyes were reddened from rage and despair. Two Marines tried to restrain him. The Prime Minister arose. He instructed them politely, "Gentlemen, please let loose the Admiral."

Miss Vectura stepped aside. The Admiral approached Morus. He leaned over the Prime Minister's desk and braced himself with clenched fists. The Prime Minister didn't feel threatened. The two men had been close friends for decades. Pullum was obviously upset. He addressed the Prime Minister by his familiar nickname. "Addie," he pleaded, "You've got to give me a squadron of ships."

Morus replied as a friend. "Ixion, what has happened?"

"Those things killed my boy. They killed Amator. The monsters attacked another convoy and blew up a merchantman. My boy fought them off until the other ships escaped. They killed him, Addie. They killed my son!"

"I'm sorry, Ixion."

"It's not pirates, or Phasmii. It's worse. They're called 'Hosticci.' The things are monsters. They're bringing hades to us all. Give me one squadron, sir. I will clear them from our space. I'll kill them all. I promise you."

"You know I can't do that, Ixion. Please, go home. Your wife needs you." Morus motioned to the marines. "Please take the Admiral home. See that he receives appropriate assistance." They expected Pullam to resist, but the Admiral was too weakened by shock and grief.

"Give me a squadron," Pullum begged. "Just one squadron!"

The Prime Minister resumed his seat. He spoke softly to his guests. "Please, excuse my friend. He speaks from a father's grief."

Morus returned his attention to Miss Vectura. Her union threatened to shut down the world's transport services. "Madam President," he began, "I understand your members' concerns about pirates. Please, tell them we're providing escorts for your merchantmen."

His remarks angered Miss Vectura. “Didn’t you hear Admiral Pullum, Prime Minister?” There are things out there attacking our ships, killing our members. You say they’re pirates? He said they’re monsters. We demand you defend us with our Navy, now!”

Mr. Inrigo leaned over his chair. His union members worked for crucial Water and Sewage Plants. “Wake up, Prime Minister!” he shouted. “The whole planet is panicked already. We need our Navy, now!”

Mr. Vis interjected. He was President of the “Planetary Power Workers Union.” He could call a strike at every power plant on Gemina. “Prime Minister,” he growled. “My children are already afraid to look up into the sky. Each morning, they ask me, ‘Daddy is the comet still coming to get us?’ They ask if the Phasmii are coming back. They cry because they’re scared. They’re terrified! I’m terrified. Now we learn about Hosticci attackers above us? The planet will fall into chaos unless you do something. Send our fleet to protect us or risk anarchy!”

In a calm voice, Morus assured his guests. “Please, be patient. We have over two years to divert Rogares. It’s true our merchant ships have been attacked by commerce raiders, but we don't need battleships to defeat pirates. We can't militarize space without risking another war. Give us more time. That's all I ask."

Vis angrily tapped his fingers on the Prime Minister's desk. "War?" He loudly questioned. "You fear war? We defeated the Phasmii with fifteen ships! Now, we have seventy-two. They're faster, more powerfully armed, better shielded than anything ever sent to space! We're not afraid of the Hosticci, or the Phasmii. They should be afraid of us!"

Miss Vectura raised her voice. "Last week, Prime Minister, we lost four merchant vessels and their crews. The *Zong* was grappled away into the expanse. Screams filled their communications, until the ship went silent. They're out there somewhere without help, and you expect our patience?"

Morus removed his glasses and rubbed his eyes. He was weary from arguing. He first addressed Mr. Inrigo. "Your union threatens to strike. You say our world is panicked? Would you worsen their situation by denying families water to drink and by closing our sewage plants?"

He then spoke to Mr. Vis. "You, sir. What will you tell your power workers when the world goes dark, and there's no energy to heat their homes, refrigerate their food, or light their cities? How will you explain why their schools and hospitals are shut down?"

He returned his attention to Miss Vectura. "Please, believe me. I understand your concerns. I'm doing everything possible. If your workers strike, there will be no transport of food, medicine, or other necessities. Be patient. Give me more time."

She replied, "Send our Navy to end Rogares and the Hosticci threat. Do so by noon tomorrow, or my union will strike."

Inrigo and Vis stepped beside Vectura. They nodded their approval. "Sir, you have until noon," warned Vis, "No later than that."

Inrigo concluded, "Send the Navy, or we all strike." They turned to leave.

Morus stood. He called to them. "Please wait. You risk our ruin. Your strikes will create utter chaos, panic, even riots! If you strike, I have no other option than to declare martial law. Military force will break your strikes. Please don't force me to do that!"

Mr. Vis warned the Prime Minister. "Be wary, Mr. Morus. We union leaders can call strikes, and we can stop them, but we can't stop riots. That's what you'll get if you impose martial law. Our world wants the Navy to defend us. You have until noon tomorrow to deploy our fleet."

The labor leaders slammed shut the Prime Minister's door as they left. He buried his face in his hands. Never would he militarize the heavens. At 12:01 pm the following day, all three unions called global strikes. Civilian power plants shut down. Transport stopped. Water supply and sewage treatment halted. The world went dark, cold, and hungry.

The Prime Minister responded by declaring martial law. He did so reluctantly. Riots erupted quickly in every City. Chaos spread worldwide. Many communities were burned to the ground. Others were abandoned. There was great loss of life and property. As Morus had been warned, the world went insane. History infamously remembered it as "Comet Day."

[End of Chapter]

www.ingramcontent.com/pod-product-compliance
Lightning Source LLC
LaVergne TN
LVHW052029200926
848481LV00005B/1624